The Open University

Mathematics: A Second Level Course

Topics in Pure Mathematics Units 1–3

SET AXIOMS

SET CONSTRUCTIONS

SETS AND NUMBERS

Prepared by the Course Team

The Open University Press

The Open University Press,
Walton Hall, Bletchley, Buckinghamshire.

First published 1973.

Copyright © 1973 The Open University.

Printed in Great Britain by
Technical Filmsetters Europe Ltd, 76 Great Bridgewater Street, Manchester M1 5JY.

SBN 335 01361 9

This text forms part of the correspondence element of an Open University Second Level Course. The complete list of units in the course is given at the end of this text.

For general availability of supporting material referred to in this text, please write to the Director of Marketing, The Open University, Walton Hall, Bletchley, Buckinghamshire.

Further information on Open University courses may be obtained from The Admissions Office, The Open University, P.O. Box 48, Bletchley, Buckinghamshire.

1.1

Unit 1   Set Axioms

"... general set theory is pretty trivial stuff really, but, if you want to be a mathematician, you need some, and here it is; read it, absorb it, and forget it."

P. R. Halmos

Contents Page

Set Books

P. R. Halmos, *Naive Set Theory*, paperback edition 1972 (Van Nostrand Reinhold).
I. N. Herstein, *Topics in Algebra*, paperback edition (Xerox College/T.A.B.S., 1964).
B. Mendelson, *Introduction to Topology*, paperback edition 1972 (Allyn and Bacon).
M. L. Minsky, *Computation: Finite and Infinite Machines*, paperback edition 1972 (Prentice-Hall).

It is essential to have these books; the course is based on them and will not make sense without them.

Unit 1 is based on **Halmos**, Sections 1–5.

Conventions

Before working through this correspondence text make sure you have read *A Guide to the Course: Topics in Pure Mathematics*.

References to the Open University Mathematics Foundation Course Units (The Open University Press, 1971) take the form *Unit M100 3, Operations and Morphisms*.

1.0   INTRODUCTION

The notion of a *set* is fundamental to mathematics. It is used either implicitly or explicitly in every mathematical discussion. It can be made to disappear, but must then be replaced by another concept which is no more fundamental. The notion of a set is usually taken to be *primitive*. That is, it is accepted as an undefined term, which everyone understands intuitively. Accompanying *set* is the idea of *set membership*, and in order to be sure that we all have a similar intuitive understanding, it is worth while writing down certain basic interrelationships between these two ideas. This is what Halmos does.

Halmos is not just doing this for the sake of interest, however. Mathematics has another aspect separate from the symbols and axioms which it discusses. This other aspect is *logic*; it is concerned basically with the ways in which statements relating to the axioms can be proved. You can think of logic as a collection of rules for manipulating the symbols by which the axioms are expressed. As an analogy, think of the axioms and theorems of mathematics as the content of a book. Logic is represented by the instructions on how to read a book—turning the pages, scanning the lines, etc.

(i) read from left to right;

(ii) . . .

. . .

. . .

content   instructions

Implicit in the logic is a demand for *consistency*. We use the fact that, if a statement and its negation can both be proved, then anything at all can be proved, and so the axiom system is inconsistent and of little interest. For example, suppose we take as our axioms all of arithmetic, together with the statement $1 = 0$. You can immediately demonstrate that every integer equals zero, and the whole arithmetic structure crumbles. We want to avoid inconsistency. Also implicit in our logic is the requirement that a statement be *either* true *or* false.

It is precisely these two demands of logic—*consistency* and the requirement that *every statement be true or false*—which force us to be very careful about using our intuition concerning the primitive notions of set and set membership. The reason for this is the existence of paradoxes. The most famous of these is known as Russell's paradox, and comes in many forms. The most blatant is the following:

let S be the set of all sets A such that A is *not* a member of A.

Linguistically, this is a perfectly good set. Suppose that this set is allowed to enter our mathematical system. Then *either*

(i) S is a member of S

or

(ii) S is not a member of S.

Assume that (i) is true. Then "S is a member of S" means, by the definition of S, that S shares the property of all members of S, namely S is not a member of S. We have "S is a member of S" implies "S is not a member of S".

Assume that (ii) is true. Then S is not a member of S, so S satisfies the condition for membership in S, namely, S is not a member of S. We have "S is not a member of S" implies "S is a member of S".

Since one of (i) and (ii) is true, we can deduce that the other is also true, and we have

"S is a member of S" and "S is not a member of S".

This is one of the dreaded inconsistencies.

It was the discovery of this paradox which caused mathematicians to examine more closely the properties of sets and set membership essential to mathematics. Certainly, if we are allowed to construct S in our mathematics, we are in trouble.

The intention of this course is not to discuss these matters in any greater detail, because they are properly the subject matter of an entire course. It is important, however, that you be exposed to an outline of one particular attempt to exclude Russell's paradox from mathematics. There are other approaches, and there is no guarantee that some other paradox will not come along. Nevertheless you will receive the flavour of the kinds of argument, and the care and precision needed in order to attack this problem.

In this unit you will meet five axioms of set construction. These axioms describe precisely how to construct new sets from old. Sets obtained using these rules are the only sets we will be allowed to talk about, and, as far as is known, no paradoxical sets can be constructed using these axioms.

The axioms themselves can be taken as either trivial or complicated. They are trivial if you understand what common everyday mathematical assumption is being described and leave it at that. They are complicated if you try to delve into the whys and wherefores of these particular axioms as opposed to others. If you question the linguistic significance of each word used, you will immediately become bogged down.

We take the view in this course that the axioms are trivial statements. Our main aim is to help you to become familiar with the basic properties and constructions of sets: set membership ($\in$), set equality ($=$), set inclusion ($\subset$), union, intersection, power set and the like. Considerable precision is necessary to define these concepts adequately, even at this "naive" level of the theory. In particular, care is needed in specifying the membership of a set.

This unit is concerned with set axioms. The next two units will show how these axioms can be used to define the basic mathematical ideas that you met in the Foundation Course, M100. Cartesian products, functions and relations all appear. We shall be able to construct the natural numbers, and, as we indicated in *Unit M100 34, Number Systems*, this permits us to construct the integers, rationals, reals and complex numbers, all based on the set theory axioms.

When you have worked through the first three units, you will have covered most of the material in the introductions to **Herstein** and **Mendelson**. Almost all mathematics texts at this level begin with a review of the basic set theory ideas that they need and with the notation that they will use; in future, you should have no difficulty in consulting another text, after first glancing through the introduction to pick up any unusual notation.

Halmos offers a challenge of a kind which is rather different from those offered in the rest of the course. Some of the material in **Herstein**, for example, can properly be considered difficult, but you will usually find that these difficulties are of a "mechanical" nature, often due to the fact that some chain of reasoning breaks down for you, because you have *forgotten* part of some previous result which is assumed in it. In **Halmos**, however, there are difficulties of a different sort: they are *intellectual* difficulties, not in following an argument, but in grasping what the argument is about.

Whether or not you accept the Halmos challenge, whether you really struggle with him or not, is up to you. What you must do is as Halmos says: "read it, absorb it, and forget it"; by "forget it" Halmos means that you should not carry the set theory precision into the mathematics of the remainder of the course. Carrying on the book analogy, once you have learned, or at least accepted, the mechanics of reading, you concentrate on the content of the book—the mathematics.

Aims

After working through this text, you should be able to:

(i) demonstrate that you have comprehended the meaning of each axiom by
distinguishing between correct and incorrect applications,
selecting those axioms needed to construct prescribed sets
(you are *not* expected to memorize the axioms);

(ii) manipulate the symbols $\in$, $\varnothing$, $\subset$, $\cap$, $\cup$, $\mathcal{P}$, $\{\}$, $=$, sufficiently well to check that $A \subset B$ or $A = B$, where A and B are sets constructed using these symbols;

(iii) appreciate the need for a deeper level of care and precision in the specification and use of the fundamental ideas of *set* and *set membership* than that of the Foundation Course.

1.1 THE AXIOM OF EXTENSION

1.1.0 Introduction

This section introduces the fundamental notions of mathematics. These are set, set membership and set equality. They will remain undefined, and we rely on intuition for their meaning. The Axiom of Extension expresses the only requirement imposed on how these three concepts are related.

The remainder of the section is concerned with the definition of *set inclusion*, and the standard properties of membership, inclusion and equality. Except for the explicit statement of the Axiom of Extension, you should be familiar with all the terms already, and so be able to concentrate on the process of carefully developing the primitive concepts into the body of mathematics that we all use. Remember that Halmos is beginning formal mathematics from scratch, but he assumes that you have mathematical sophistication.

READ Halmos: the Preface.

Note especially *page v, lines − 9 to − 7.*

1.1.1 The Axiom

READ Halmos: Section 1, pages 1 to 3.

Notes

(i) *Halmos: page 2, line 7.*
$x \in A$ should be read as *x is a member of the set A* or *x belongs to A*. The word "contained" should be reserved for another use.

(ii) *Halmos: page 2, line − 8 (The Axiom).*
You have met the phrase if and only if before (in *Unit M100 17, Logic II*), but it takes some getting used to. It is the most common way of saying

is implied by, and implies or, symbolically, ⇔.

Wherever you see

A if and only if *B*   or $A \Leftrightarrow B$,

you are being given two pieces of information: first, that

A is implied by *B*   i.e.   *A* if *B*   i.e. $A \Leftarrow B$,

and second that

A implies *B*   i.e.   *A* only if *B*   i.e. $A \Rightarrow B$.

Thus, whenever you know *A* is true, you can conclude that *B* is also true, and vice versa.

The Axiom of Extension gets its name because it assures us that if (and only if) two sets with different names have the same extent, that is, contain exactly the same elements or members, then they are equal. Of course, we treat equal sets as being indistinguishable from each other. This axiom guarantees that we are all talking about the same set when we specify a set using further axioms.

(iii) *Halmos: page 3, lines 1 to 8.*
Halmos is demonstrating an interpretation of the primitive terms "set" and "∈" for which the Axiom of Extension does not hold. Thus he demonstrates that the Axiom of Extension really does say something about the relationship between "set", "∈" and equality.

(iv) **Halmos** : *page 3, line 11.*

Halmos uses $A \subset B$ to mean *A is included in or equal to B*. This is not the same use as in *Unit M100 1, Functions,* but it agrees with **Herstein**, and will be used throughout this course. "$A \subset B$" is usually read as "*A* is contained in *B*", but there is a potential source of confusion. Halmos sometimes translates "$x \in A$" as "*x* is contained in (is a member of) *A*", and "$X \in A$" as "*X* is contained in (is a subset of) *A*". To avoid confusion, you should read "$x \in A$" as "*x* belongs to *A*" or "*x* is a member of *A*" and reserve "contained" for set inclusion ($\subset$).

(v) **Halmos** : *page 3, line − 12.*

This is an extremely important technique. You will have a chance to apply it in Section 1.4.

(vi) **Halmos** : *page 3, line − 4.*

Inclusion is *transitive* because $A \subset B$ and $B \subset C$ implies $A \subset C$, but belonging is not, because $a \in x$ and $x \in A$ does not permit us to deduce that $a \in A$. For example, we can regard a book as a set of words, and a library as a set of books. A library, under our definition, is not then a set of words. Thus a particular word may be a member of a book which is itself a member of a library, but the word itself is not a member of the library.

Remark

There are no examples or exercises in this section because so far we do not even know that there are any sets. Halmos has used examples such as a pack of wolves simply as an intuitive basis for the idea of a set. The one axiom that he has formulated does not enable him to construct any sets. Later he will assume that certain sets exist.

1.2 THE AXIOM OF SPECIFICATION

1.2.0 Introduction

The Axiom of Specification is the major principle of set theory. It is designed specifically to circumvent Russell's paradox (see *Unit M100 17, Logic II*, section 17.1.2) which arises from the use of the concept of the "set of all sets". To avoid this paradox, the axiom says that, if you have a set A, and if you can state some condition on the elements in A, in some acceptable language, then those elements in A which satisfy the condition do indeed form a set.

The axiom is used in two ways:

(i)   to guarantee that certain "things" really are sets;

(ii) to give names to, or describe, sets in terms of properties possessed by their elements.

For example, if you have a set A of red objects, some of which are round, and some square, then the Axiom of Specification permits you to select the round red objects from A, and form a set of them. You then know (i) that they really do form a set and (ii) that the new set has precisely the round red objects from A as its members, and no others.

All this may seem needlessly pedantic, but it is necessary in order to avoid trouble. For example, we define a set X to be *t-circular* if there exist sets $x_1, x_2, \ldots, x_t$ such that

$$X \in x_1 \in x_2 \in \ldots \in x_t \in X.$$

We call a set *circular* if it is t-circular for some t; otherwise it is *non-circular*. Now let

$$N = \{x : x \text{ is a non-circular set}\}.$$

Certainly if $N \in N$, then N is non-circular, yet $N \in N \in N$ so N is 1-circular, a clear contradiction. Also if $N \notin N$, then N is circular, and so $N \in x_1 \in \ldots \in x_t \in N$. But then $x_t \in N$ so x_t cannot be circular, and yet

$$x_t \in N \in x_1 \in \ldots \in x_{t-1} \in x_t,$$

so x_t is t-circular, which gives another contradiction. We are in trouble if we admit N as a set! N is excluded because its members are not required to belong to a previously defined set A. If we take an arbitrary set A, and put

$$N = \{x \in A : x \text{ is a non-circular set}\},$$

then we have a valid application of the Axiom of Specification. If we now try to reach a contradiction, we cannot conclude from $N \notin N$ that N is circular, since N may fail to be a member of N by virtue of $N \notin A$.

The difficulties seem to arise because the language we are using is rich enough to permit self-describing sentences. We are forced to impose some restriction on our freedom to specify sets by means of sentences. We could spend an entire course exploring these difficulties, but for our present purposes the elementary description in this section will be adequate to provide the flavour.

The definition of a sentence in this section is the first occurrence of a *recursive definition*. A recursive definition is basically a definition in the form of a set of instructions about how to build up more and more complex examples from simple ones. It is a fundamental technique in mathematics; we shall discuss it explicitly later, when studying **Minsky**.

1.2.1   Sentences

READ **Halmos**: page 4, line 1 to page 5, line 2.

Notes

(i)   **Halmos**: page 4, line − 10.
$\{x \in A : x \text{ is married}\}$ is read "the set of all x in A such that x is married".

(ii) **Halmos**: page 4, line − 5.
This is incorrect. See *Genesis*, Chapter 4, verse 25 to Chapter 5, verse 4.

General Comment

The difficulty in the previous passage is that we have no formal definition of a *sentence*. You may find it an interesting exercise to try to define what we mean by a sentence in English. Do not spend too long on it though, because it has given a lot of linguists a good deal of trouble. Fortunately in mathematics the job is relatively easy. Instead of trying to identify all possible sentences (an impossible task, since there are not finitely many), we specify how to construct all possible sentences by building them up from smaller ones. The smallest, or starting, sentences are called atomic sentences. For example, $a \in x$, $b \in B$, $X = Y$ are atomic sentences. We then use the logical connectives (which we introduced in *Unit M100 11, Logic I*) such as not, or, and, implies, if and only if, and the quantifiers, there exists and for all (which you met in *Unit M100 17, Logic II*) to build up more complicated sentences. You should not attach particular importance to the next reading passage, except to observe how this definition works.

READ **Halmos**: page 5, line 3 to page 6, line 4.

Notes

(i)   **Halmos**: page 5, line − 15.
Using just this rule, and the atomic sentence $x \in A$, we can build up

$$(\text{not } x \in A)$$
$$(\text{not}(\text{not } x \in A))$$
$$(\text{not}(\text{not}(\text{not } x \in A)))$$
etc.

(ii) **Halmos**: page 5, line − 6.
Alternatively, replace the dashes in "—implies—" by sentences and enclose the result in parentheses.

(iii) **Halmos**: page 5, line − 1.
Halmos has contradicted his convention here because, strictly speaking, $(x \in A)$ is not a sentence. (The parentheses should not be there.) In fact we insert and delete parentheses to make sentences readable and unambiguous. There is no need to be pedantic about this matter of parentheses. In particular, we shall frequently write

for all $x \in A$ or $\forall x \in A$

to mean

(for all x $x \in A$)   or $(\forall x \quad x \in A)$.

Summary

To obtain a new sentence, perform one of the following constructions on arbitrary old sentences S and T and an arbitrary letter x.

(i)   (not S)

(ii) (S or T)   (S and T)

(iii) (S if and only if T)

(S implies T), in other words, (if S, then T)

(iv) (for some x S), in other words, $(\exists x$ $S)$
(for all x S), in other words, $(\forall x$ $S)$.

Note that a sentence need not "make sense" when interpreted. For example, $(\forall x(\forall y(x = y \Rightarrow x \neq y)))$ is a perfectly good sentence.

In *Unit M100 11, Logic I*, we discussed the problem of deciding when two sentences are the same. This notion of syntax is, of course, extremely important, but would lead us too far astray at this point. We shall stick to our intuitive notions and assume that, for example,

(not(not $x \in A$)) is the same as $x \in A$.

Example 1

Show how the sentence

(for all x(for all y(for all z($x \in y$ and (not $y \in z$)) implies $x \in z$)))

can be built up from atomic sentences using the rules.

Solution 1

$x \in y$, $y \in z$, $x \in z$ are atomic sentences,

so, from rule (i),

(not $y \in z$) is a sentence.

From rule (ii),

($x \in y$ and (not $y \in z$)) is a sentence.

From rule (iii), it follows that

(($x \in y$ and (not $y \in z$)) implies $x \in z$) is a sentence.

Three applications of rule (iv) will produce the entire sentence.

Example 2

For each of the following, state whether or not it is a sentence, and if not, why not.

(a) For all x $y \in A$.
(b) For all ($x \in A$ and $y \in B$).
(c) A implies $x \in A$.

Solution 2

(a) is a sentence.
(b) breaks rule (iv) and (c) breaks rule (iii), so these are not sentences.

SAQ 1★

State which of the following are sentences according to the definition in **Halmos**. (Faults do not lie with the parentheses.)

(a) (if $x \in A$ then (not $x \in A$)).
(b) ($x \in A$ and $y \in B$).
(c) (for some x (for all x $x \in A$)).
(d) (not x).
(e) (x or y) $\in A$.

(Solution is given on p. 25.)

★ SAQ stands for Self-Assessment Question. These questions are numbered sequentially throughout the text. Solutions are given in the final section, on the pages indicated. For the purpose of these questions please see *A Guide to the Course: Topics in Pure Mathematics*.

SAQ 2

Show how to build up the following sentence, stating the rule used at each stage.

$$(\exists x((\forall y \quad y \in A) \text{ implies } ((\exists x \quad x \in y) \text{ or } x \in A))).$$

(Solution is given on p. 25.)

1.2.2 The Axiom

READ Halmos: page 6, line 5 as far as the statement $B = \{x \in A : S(x)\}$.

Example

In each of the following sentences, the occurrence of x which is free is in the atomic sentence $x \in A$. All other occurrences of x are not free (i.e. are bound).

$x \in A$

(for some x $\quad x \in B$) and $x \in A$ $\qquad$ $(\exists x \quad x \in B)$ and $x \in A$

(for all $x(x \in B$ implies $y \in x))$ or $(x \in A)$ $\qquad$ $(\forall x(x \in B \Rightarrow y \in x))$ or $(x \in A)$.

General Comment

This passage is important because it states precisely the *only* way we have (so far) of specifying a new set B starting with a given set A. For example, we might try to construct a set as follows—it is *not* a valid use of the Axiom of Specification:

$$B = \{x : \forall a(a \in x \Rightarrow a \in A)\}.$$

B is intuitively the collection of subsets of A, since each x has only members of A as members. The sentence

$$S(x) = \forall a(a \in x \Rightarrow a \in A)$$

is a perfectly good sentence, with x free. However, x is not constrained to belong to any previously given set, and that is an essential requirement in the Axiom of Specification. (See also our example on p. 10.) To ensure that such sophisticated ideas as the collection of subsets of the set A actually does form a set, we shall have to introduce further axioms later (Section 1.4).

to use the Axiom of Specification you need:

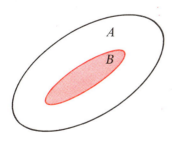

a set A of which B
is to be a subset

a condition which distinguishes the
elements of B from the remainder of A,
in the form of an acceptable sentence

1.2.3 Application of the Axiom

The next reading passage concerns the intuitive idea of the *universe of discourse*, which refers to the totality of objects we wish to talk about in mathematics. Halmos concludes, using just our two axioms, that such a universe of discourse cannot be a set. If it were a set, we could reach a contradiction!

It is this conclusion which justifies the care we have taken with the rules on how to define sets. If you do not take this care, you might allow the universe of discourse to be a set, and you would then fall directly into Russell's paradox. (See Section 1.0.) Since we want our mathematics to be free from contradictions, the care we have been taking is necessary★.

READ Halmos: page 6, line 17, to the end of the section on page 7.

Notes

(i) **Halmos**: *page 6, line − 5.*
This proof makes use of the unstated assumption:

if A is a set, then, for all x, *either*[†] $x \in A$ or $x \notin A$.

(ii) **Halmos**: *page 6, line − 1.*
A better statement is

"No set contains everything."

(iii) **Halmos**: *page 7, line 2.*
Halmos is saying "no universe of discourse can be a set" or "there is no universal set".

(iv) **Halmos**: *page 7, line − 2.*
Underline *have at hand a set* as it is extremely important.

★ but may not be sufficient . . .

† We shall use italic type for "*either . . . or . . .*" when we mean "but not both". We shall use roman type for "either . . . or . . ." when we mean "either . . . or . . . or both" (see **Halmos**: page 5, line 17).

1.3 UNORDERED PAIRS

1.3.1 The Empty Set

READ **Halmos**: page 8.

Notes

(i) **Halmos**: page 8, line 9.

$x \neq x$ is a sentence because it is a short form for (not $x = x$).

(ii) **Halmos**: page 8, line 11.

This is a typical use of the Axiom of Extension. Once we have defined the members of a set, we know there is a *unique* set having just those members. For example, a set of apples which has no apples (no members) is the *same set* as a set of oranges which has no oranges (no members). Both are *the* empty set.

(iii) **Halmos**: page 8, lines −5 and −4.

This is not only valuable advice; it is the only way that you can proceed, because of the nature of the empty set. For example, in the previous note we examined a set of apples and a set of oranges. To see that they are the same set, let us suppose that they are different. Then the Axiom of Extension says that one set must have a member that the other does not, and this cannot happen. This is a good example of how the Axiom of Extension is used to show that two sets, which are arrived at by different means (apples, oranges) and which may have been given different names, are nevertheless indistinguishable, and so are the same set.

SAQ 3

(i) Does $\emptyset \in \emptyset$?

(ii) Show that $A \subset \emptyset \Rightarrow A = \emptyset$.

(Solution is given on p. 25.)

1.3.2 Pairing

READ **Halmos** : page 9, line 1 to page 10, line 7 (end of paragraph).

Notes

(i) **Halmos**: page 9, line − 17.

The Axiom of Pairing, as stated, guarantees only that there is some set A containing *at least* the sets a and b as members, but there may be other elements in A as well. The Axiom of Specification then permits us to identify a subset of A containing just a and b, and the Axiom of Extension guarantees that there is only one such set. This is typical of the way in which these two axioms are used in conjunction with the other axioms.

(ii) **Halmos**: page 9, line − 4.

Notice that this version of the axiom incorporates the above-mentioned application of the Axiom of Specification. It asserts the existence of a set which contains only the sets a and b as members.

READ **Halmos**: page 10, line 8 to the end of the section on page 11.

(The solution to the exercise on *page 10* is given in Note (ii) which follows.)

Notes

(i) **Halmos**: page 10, line 11.

The term *singleton* is frequently used. It is closely allied to the set-theoretic definition of the number 1, which is the basis of the number system. Notice that $\{a\}$ is an example of the convention that distinct members of a set are written only once. Notice particularly that $\emptyset \neq \{\emptyset\}$. The set $\{\emptyset\}$ has a member, namely $\emptyset$, whereas $\emptyset$ itself has no members.

(ii) ***Halmos***: *page 10, the exercise on line* -13.
The sets obtained are all distinct. Intuitively, each new set is a pair of old sets. If two sets turned out to be equal, they would have been constructed from the same pair of old sets. A more careful proof would involve mathematical induction. The problem is not worth spending that much time on, however.

(iii) ***Halmos***: *page 10, line* -4.
For some, but not all, sentences $S(x)$, $\{x : S(x)\}$ is a valid set. For example, $x = x$ is a sentence, with x free, but $\{x : x = x\}$ is not a set because x is not confined to a previously specified set. The Axiom of Specification is satisfied if $S(x)$ includes a restriction on x to be a member of a previously defined set. (See also our example on page 10.)

(iv) ***Halmos***: *page 10, line* -3.
"In case" is an alternative form of "If".

(v) ***Halmos***: *page 10, line* -3 *and page 11, line 1.*
Underline the third *is* for emphasis.

(vi) ***Halmos***: *page 11, line 7.*
Insert *However* at the beginning of the sentence for emphasis.

(vii) ***Halmos***: *page 11, line 9.*
Delete "the" and "about" and send all contributions to the Rest Home for tired Deans!

Example

Show that $A = \big\{\{\emptyset\}, \{\{\emptyset\}\}\big\}$ is a set.

Solution

To show that A is a set, we use the Axiom of Pairing three times. We know that $\emptyset$ is a set, and by the Axiom of Pairing applied to $\emptyset$ and $\emptyset$, $\{\emptyset\}$ is a set. (We have used the Axiom of Specification here as well to ensure that there *is* a set containing just $\emptyset$ as a member.) Repeating the process with $\{\emptyset\}$, $\{\{\emptyset\}\}$ is a set. We now apply pairing to $\{\emptyset\}$ and $\{\{\emptyset\}\}$ and thus establish that

$$\big\{\{\emptyset\}, \{\{\emptyset\}\}\big\}$$

is a set.

SAQ 4

Let $A = \big\{\{\emptyset\}, \{\{\emptyset\}\}\big\}$. Show that there exists a set x such that $x \in A$ and $x \subset A$.

(Solution is given on p. 25.)

SAQ 5

Suppose a, b, c are distinct sets. Can you assert that there exists a set containing just a, b, c as members?

(Solution is given on p. 25.)

1.4 UNIONS AND INTERSECTIONS

There is virtually nothing new in this section. It provides you with practice in sorting through the symbolism which expresses ideas you met in the Foundation Course.

1.4.1 Unions

READ Halmos: page12, line 1 to page14, line 15.

Notes

(i) *Halmos: page 12, line − 8.*
Except in the few cases in which no confusion can arise, you should always put quantifiers ($\exists x, \forall x$) *in front of* the free occurrence of x to which they apply, and write the quantifiers in the appropriate order. Otherwise you may create ambiguous sentences. For example, if x, y are interpreted as real numbers,

$$\exists y (y \geqslant x) \forall x$$

is ambiguous because it could mean *either*

$$\forall x \exists y (y \geqslant x),$$

which is true for real numbers, *or*

$$\exists y \forall x (y \geqslant x)$$

which is false.

(ii) *Halmos: page 13, lines 12 and 14.*
See SAQ 6, parts (i) and (ii).

(iii) *Halmos: page 13, line − 5.*
This statement is established in Example 1 below.

(iv) *Halmos: page 13, line −1.*
See SAQ 6, part (iii).

(v) *Halmos: page 14, line 10.*
"it is easy to prove" means "the method of proof should be obvious and the proof is straightforward".

Example 1

Show that $A \cup \varnothing = A$.

Solution 1

We proceed in two steps. We show first that

$$A \cup \varnothing \subset A,$$

and, second, that

$$A \subset A \cup \varnothing.$$

(See *Halmos: page 3, line − 12.*) To prove the first statement, we show that,

if $x \in A \cup \varnothing$, then $x \in A$.

Now $x \in A \cup \varnothing$ implies that either $x \in A$ or $x \in \varnothing$ (Axiom of Unions).

Since $x \in \varnothing$ is impossible, $x \in A \cup \varnothing$ tells us that $x \in A$.

To prove the second statement, we show that,

if $x \in A$, then $x \in A \cup \varnothing$.

This is clear from the definition of $A \cup \varnothing$, since if $x \in A$, then certainly $x \in A$ or $x \in \varnothing$.

Example 2

Show that $x \in \mathscr{C}$ implies $x \subset \bigcup \mathscr{C}$.

Solution 2

To show that $x \subset \bigcup \mathscr{C}$, we let y be an arbitrary member of x. Then $y \in x$ implies that $y \in \bigcup \mathscr{C}$ by the definition of $\bigcup \mathscr{C}$. Hence $x \subset \bigcup \mathscr{C}$.

SAQ 6

Halmos says that every student of mathematics should prove the various statements about unions of sets. At this point you should not let the word *prove* bother you. Halmos means that, using the definitions of the terms such as $\bigcup, \varnothing, \{A\}$ etc., you should be able to verify that the statements are correct.

(i) Show that $\bigcup \{X : X \in \varnothing\} = \varnothing$.

(ii) Show that $\bigcup \{X : X \in \{A\}\} = A$.
(HINT: Use the method of Example 1 above.)

(iii) Show that $A \subset B$ if and only if $A \cup B = B$.
(HINT: There are two things to show in an "if and only if" statement. See Note (ii) of Section 1.1.1.)

(Solution is given on p. 26.)

1.4.2 Intersections

*READ **Halmos**: page 14, line 16 to the end of the section on page 16.*

Notes

(i) ***Halmos**: page 14, line − 5.*
You should verify these statements if you do not recognize them as being obviously true.

(ii) ***Halmos**: page 15, line 10.*
A diagram may help:

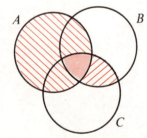

$A \cup (B \cap C)$ is represented by diagonal stripes and pink shading

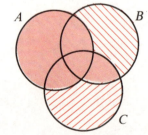

$(A \cup B) \cap (A \cup C)$ is represented by pink shading only

SAQ 7

Complete the exercise in ***Halmos**: page 16*. The phrase "necessary and sufficient" is equivalent to "if and only if". You are being asked to show that

$$(A \cap B) \cup C = A \cap (B \cup C) \text{ if and only if } C \subset A.$$

Remember, you have two things to do!

(HINT: Use the statements on ***Halmos**: page 15*.)

(Solution is given on p. 27.)

1.5   COMPLEMENTS AND POWERS

1.5.0   Introduction

This section is about subsets. Halmos introduces a fair number of mathematical statements concerning the relationships between set differences $(A - B)$, unions and intersections. There is no need to commit these statements to memory or otherwise endow them with special importance. When we begin to use them extensively, you will quickly become familiar with them. You should understand them sufficiently so that, if needed, you can check them, or work them out.

1.5.1   Complements

*READ **Halmos**: page 17, line 1 to page 19, line 17.*

Notes

(i) **Halmos**: *page 17, line 4.*
Underline *not* for emphasis.

(ii) **Halmos**: *page 17, line − 6.*
Draw diagrams to illustrate this statement.

(iii) **Halmos**: *page 18, line 14.*
The symmetric difference of two sets is a useful concept but not so frequently referred to as the concepts introduced in previous sections.

(iv) **Halmos**: *page 18, lines − 17 to − 15.*
Notice that here we have an example of a *group*. (See *Unit M100 30, Groups I*. The set is the set of subsets of some set E; the binary operation is symmetric difference.)

(v) **Halmos**: *page 18, last paragraph, to page 19, line 17.*
You will probably find it useful to reread this passage from **Halmos**. Many authors slide over this point, with varying degrees of inaccuracy, and it is worth while getting it straight, and then not worrying about it.

1.5.2 The Power Set

READ *Halmos*: *page 19, line 18 to the end of the section on page 21.*
(The exercise appears as an example below.)

Note

Halmos: *page 20, line 7.*
This is a straightforward exercise in the method of proof by mathematical induction.
(See *Unit M100 17, Logic II*.) We shall say a lot more about finite and infinite sets
later in the course.

Example

(See *Halmos*: *page 20, line − 5.*)

Show that, if E and F are sets, then

(i) $\mathscr{P}(E) \cap \mathscr{P}(F) = \mathscr{P}(E \cap F)$.

(ii) $\mathscr{P}(E) \cup \mathscr{P}(F) \subset \mathscr{P}(E \cup F)$.

Solution

(i)   We are really being asked to demonstrate that the following is a commutative
diagram.

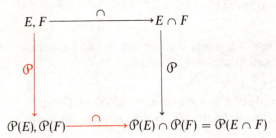

We have

$$A \in \mathscr{P}(E) \cap \mathscr{P}(F) \Leftrightarrow A \in \mathscr{P}(E) \text{ and } A \in \mathscr{P}(F)$$

$$\Leftrightarrow A \subset E \text{ and } A \subset F$$

$$\Leftrightarrow A \subset (E \cap F)$$

$$\Leftrightarrow A \in \mathscr{P}(E \cap F).$$

Therefore, by the Axiom of Extension,

$$\mathscr{P}(E) \cap \mathscr{P}(F) = \mathscr{P}(E \cap F).$$

(ii) The argument is very similar.
Let $A \in \mathscr{P}(E) \cup \mathscr{P}(F)$.

Then $A \in \mathscr{P}(E)$ or $A \in \mathscr{P}(F)$.

This means that $A \subset E$ or $A \subset F$ which implies that $A \subset E \cup F$. From this we
find that $A \in \mathscr{P}(E \cup F)$ and so

$$\mathscr{P}(E) \cup \mathscr{P}(F) \subset \mathscr{P}(E \cup F).$$

The sets $\mathscr{P}(E) \cup \mathscr{P}(F)$ and $\mathscr{P}(E \cup F)$ are not in general equal, however. For example,
let $E = \{\varnothing\}, F = \{\{\varnothing\}\}$; then $E \cup F = \{\varnothing, \{\varnothing\}\}$. We have

$$\mathscr{P}(E) = \{\varnothing, \{\varnothing\}\} \quad , \quad \mathscr{P}(F) = \{\varnothing, \{\{\varnothing\}\}\}$$

and

$$\mathscr{P}(E \cup F) = \{\varnothing, \{\varnothing\}, \{\{\varnothing\}\}, \{\varnothing, \{\varnothing\}\}\},$$

whereas

$$\mathscr{P}(E) \cup \mathscr{P}(F) = \{\varnothing, \{\varnothing\}, \{\{\varnothing\}\}\}.$$

SAQ 8

(**Halmos**: *page 21*)

Let E be a collection of sets.

(i)   Show that $\bigcap_{X \in \mathcal{P}(E)} X = \varnothing$.

(ii) Show that $E = \bigcup \mathcal{P}(E)$.

(iii) Show that $E \subset \mathcal{P}(\bigcup E)$.

(Solution is given on p. 28.)

1.6 SUMMARY OF THE TEXT

In this unit we have examined the five axioms of set construction listed below.

Extension: Two sets are equal if and only if they have the same elements.

Specification: To every set A, and to every condition $S(x)$, there corresponds a set B whose elements are exactly those elements x of A for which $S(x)$ holds.

Pairing: For any two sets, there exists a set that they both belong to.

Unions: For every collection of sets, there exists a set that contains all the elements that belong to at least one set of the given collection.

Powers: For each set, there exists a collection of sets that contains among its elements all the subsets of a given set.

These are the basic axioms of **Halmos**' set theory. (Intersections did not need an axiom.) There is also the (temporary) working assumption that *a set exists*, since otherwise the theory would be vacuous. Together with this working assumption, the axioms constitute a complete description of how sets may be constructed.

In the process of examining these axioms, we have introduced the notations of

membership	$\in$
inclusion	$\subset$
intersection	$\cap$
union	$\cup$
singleton	$\{a\}$
power set	$\wp$
set complement	$A - B$

These form the basic language of mathematics, and you should be conversant with them.

Herstein: *pages 2–5* and **Mendelson**: *pages 2–8* duplicate some of this material, but in a more succinct form. The significant differences are listed below.

	Halmos	**Herstein**	**Mendelson**
set difference	$A - B$	$A - B$	$C_A(B), A/B, A - B$
complement (relative to a fixed set E)	A'	$E - A$	$C(A)$
improper subset of A	A	A	$A, \varnothing$

We have also explored three important techniques for proving statements about sets:

(i) To show that $A \subset B$, show that

$\forall x(x \in A \text{ implies } x \in B)$.

(ii) To show that $A = B$, show that

$A \subset B$ and $B \subset A$.

(iii) To prove a statement about the empty set $\varnothing$, prove that the negation is false.

1.7 FURTHER SELF-ASSESSMENT QUESTIONS

SAQ 9

Describe briefly the following sets. (*A* is a set.)

(i) $\{p : p \in A \text{ and } p \neq p\}$.

(ii) $\{x : x \in A\}$.

(iii) $\{x : x \in y \text{ and } y \in B\}$, where $B = \{\varnothing, \{\varnothing\}, \{\{\varnothing\}\}\}$.

(iv) $\{x : x \in A \text{ or } x = \varnothing\}$.

(Solution is given on p. 28.)

SAQ 10

State whether or not the following are valid applications of the Axiom of Specification. If not, state why not. In (ii), *A* and *B* are known sets.

(i) $\{x : x = x\}$.

(ii) $\{x : x \in A \text{ and } x \in B\}$.

(Solution is given on p. 28.)

SAQ 11

State whether the following statements are

> always true
> always false
> sometimes true—state when.

A is a set.

(i) $A \cup \{\varnothing\} = A$.

(ii) $\mathcal{P}(A) = A$.

(iii) $A \in \{A\}$.

(Solution is given on p. 28.)

SAQ 12

In **Mendelson**: *page 3, line 14*, the definition of a *proper subset* of the set *A* is different from the definition in **Halmos**. *A* itself, and $\varnothing$, are *improper*, and all other subsets of *A* are *proper*.

Describe those sets which have no (Mendelson) proper subset.

What axioms do you need to use to assert that, for a given set *E*, the collection of subsets of *E* which themselves have no (Mendelson) proper subset, is indeed a set?

(Solution is given on p. 29.)

1.8 POSTSCRIPT

Letter to the Editor of *The Mathematics Teacher*,

DEAR EDITOR,

In your February 1968 issue, on page 122, a writer describes an attempt to prove to the satisfaction of high school students that every set includes the empty set. I believe that in mathematical teaching, a healthy dose of enlightened anti-intellectualism is in order. Specifically I recommend this approach: DEFINITION. Let A, B be nonempty sets; then $A \subset B$ means every member of A is in B. We also *define* $\varnothing \subset B$ for all B. This is no question of proof — this is a definition. Tell the students, if they feel unhappy with this definition, to wait a few moments until you can prove your first theorem: THEOREM. $A \cap B \subset B$. *Proof. Case* 1. A and B meet (easy). *Case* 2. A and B disjoint (true by the definition). Thus the definition was adopted because mathematicians hate exceptions to theorems — or multiplying hypotheses. (THEOREM. If A and B meet, $A \cap B \subset B$.)

Note that the definition just given is not redundant — only a little inelegant — and maybe not so inelegant! It's exactly like: DEFINITION. For positive integer $n, n! = 1, 2, \ldots n$; and $0! = 1$. THEOREM. $n! = n[(n - 1)!]$. *Proof. Case* 1, $n > 1$ (easy). *Case* 2. $n = 1$ (true by the definition). Elegance is a matter of taste, and extracting facts hidden in a mathematical definition by manipulations with empty sets, empty unions and empty intersections is silly bombast for many high school students. The general public is also likely to be turned off by these antics. On leaving for a picnic, my wife asked me if I had brought all the soda from the refrigerator. I said yes. Later at the picnic she asked for the soda. I said: "There wasn't any there." Then again, one of my colleagues was asked about his blessed event: "Was it a boy or a girl?" He answered, "Yes." My secretary, who once had a course in quadratic equations, said, "There are two equal students waiting to see you." As a final example: One of my mathematical friends had to take a sanity test. The lawyer asked him, "Professor, if Rome is the capital of New York, must the moon be made of green cheese?" "Of course," he answered brightly. "Lock him up!" said the judge.

ALBERT WILANSKY
Lehigh University
Bethlehem, Pennsylvania

1.9 SOLUTIONS TO SELF-ASSESSMENT QUESTIONS

Solution to SAQ 1

(a), (b) and (c) are sentences.
(d) breaks rule (i) and (e) breaks rule (ii), so these are not sentences.

Solution to SAQ 2

Begin with the atomic sentences. The parentheses tell you how to break down the sentence.

$y \in A$, $x \in y$, $x \in A$	atomic sentences
$(\forall y \quad y \in A)$, $(\exists x \quad x \in y)$, $x \in A$	rule (iv)
$((\exists x \quad x \in y)$ or $x \in A)$	rule (ii)
$((\forall y \quad y \in A)$ implies $((\exists x \quad x \in y)$ or $x \in A))$	rule (iii)
$(\exists x \quad ((\forall y \quad y \in A)$ implies $((\exists x \quad x \in y)$ or $x \in A)))$	rule (iv)

Notice that the *last* "$\exists x$" *to be added* governs only the atomic sentence $x \in A$. $x \in y$ is governed by the earlier "$\exists x$". Thus the following sentence has exactly the same meaning as the one given:

$$(\exists x((\forall y \quad y \in A) \text{ implies } ((\exists z \quad z \in y) \text{ or } x \in A))).$$

Solution to SAQ 3

(i) $\varnothing$ has no members, so $\varnothing \notin \varnothing$.

(ii) To show that $A \subset \varnothing \Rightarrow A = \varnothing$, we assume that $A \subset \varnothing$ and deduce that $A = \varnothing$.

   If $A \subset \varnothing$, then for all $x \in A$, x must be a member of $\varnothing$. But $\varnothing$ has no members. Thus A can have no members either, so $A = \varnothing$.

   Alternatively, it is always true that $\varnothing \subset A$. If it is also true that $A \subset \varnothing$, then immediately $A = \varnothing$. (See **Halmos**: *page 3, line* -12.)

Solution to SAQ 4

Put $x = \{\{\varnothing\}\}$. (The only other choice for x is $\{\varnothing\}$, which does not work.) $x \in A$ is clear by the specification of A. But $x \subset A$ since x has only $\{\varnothing\}$ as a member, and this is a member of A as well.

Solution to SAQ 5

No. The Axiom of Pairing will only guarantee the existence of sets with one or two distinct members. We need a further axiom to do more. We can construct the sets

$$\{a, \{b, c\}\}, \{\{a, b\}, c\}$$

and the like, but a, b, c are not all members of any of these.

Solution to SAQ 6

(i) Go back to the definition of $\bigcup$. (See **Halmos***: page 12*.) An element x can be a member of $\bigcup\{X : X \in \varnothing\}$ only if, for some X, $x \in X$ and $X \in \varnothing$.

Since X cannot be a member of $\varnothing$, there can be no members of $\bigcup\{X : X \in \varnothing\}$, whence

$$\bigcup\{X : X \in \varnothing\} = \varnothing.$$

(ii) Let $P = \bigcup\{X : X \in \{A\}\}$. We shall show first that $P \subset A$ and then that $A \subset P$. We shall then be able to conclude that $A = P$.

To show that $P \subset A$, we need only show that if $p \in P$, then $p \in A$.

Let $p \in P$, then by the definition of $\bigcup$, for some $X \in \{A\}$, $p \in X$.

But A is the unique member of $\{A\}$. Thus $p \in A$.

We have shown (laboriously), that, if $p \in P$, then $p \in A$, which means that $P \subset A$.

Similarly, to show that $A \subset P$, let $x \in A$. Then $x \in A$ and $A \in \{A\}$, so $x \in P$. Consequently $A \subset P$.

We have now shown that $P \subset A$ and $A \subset P$, and so $A = P$. It is not necessary to be so verbose—we have provided a full answer so that you can see the thought processes which go into making up a proof.

(iii) We shall first show that $A \subset B$ implies $A \cup B = B$ (the "only if" part). The statement $B \subset A \cup B$ is always true, so we need only show that

$$A \subset B \Rightarrow A \cup B \subset B.$$

Now for any $x \in A \cup B$, either $x \in A$ or $x \in B$.

If $A \subset B$, then $x \in A$ implies $x \in B$.

Thus, in both cases, if $x \in A \cup B$, then $x \in B$, whence $A \cup B \subset B$.

For the "if" part, we assume that $A \cup B = B$ and show that $A \subset B$. To do this we let x be an arbitrary member of A. We want to show that $x \in B$.

Now if $x \in A$, then $x \in A \cup B$.

But $A \cup B = B$, so if $x \in A$ then $x \in B$, whence $A \subset B$.

Our proof is now complete.

Solution to SAQ 7

Although diagrams can never constitute a proof, they can help to focus attention on the right attack.

Here we have:

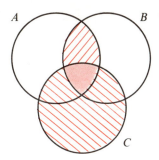

$(A \cap B) \cup C$ is represented by
diagonal stripes and pink shading

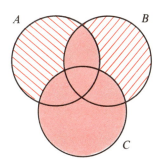

$(A \cup C) \cap (B \cup C)$ is represented by
pink shading only

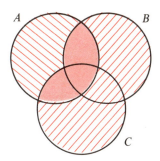

$A \cap (B \cup C)$ is represented by
pink shading only

We first assume that $(A \cap B) \cup C = A \cap (B \cup C)$ and show that $C \subset A$.

From the above equation, we can conclude that $(A \cap B) \cup C \subset A$ (and also $\subset B \cup C$). Now $C \subset (A \cap B) \cup C$, so $C \subset A$. (Alternatively, you can show that $x \in C$ implies $x \in A$.)

Secondly, we assume that $C \subset A$, and show that $(A \cap B) \cup C = A \cap (B \cup C)$. We use the second distributive law, which states

$$(A \cap B) \cup C = (A \cup C) \cap (B \cup C).$$

(See **Halmos**: *page 15, line 9, and page 13, line* -4.)

Now, if $C \subset A$, then $A \cup C = A$.

(See **Halmos**: *page 13, line* -1.)

Thus $(A \cap B) \cup C = A \cap (B \cup C)$.

Our proof is now complete.

Solution to SAQ 8

(i) Let $A \in \mathcal{P}(E)$. Then $\bigcap_{X \in \mathcal{P}(E)} X \subset A$ since $\mathcal{P}(E) \neq \varnothing$.

In particular, $\varnothing \in \mathcal{P}(E)$, so $\bigcap_{X \in \mathcal{P}(E)} X \subset \varnothing$.

But $\varnothing$ has only one subset, namely itself. Thus $\bigcap_{X \in \mathcal{P}(E)} X = \varnothing$.

Alternatively, suppose $y \in \bigcap_{X \in \mathcal{P}(E)} X$.

Then for any $X \in \mathcal{P}(E)$, $y \in X$. In particular, $\varnothing \in \mathcal{P}(E)$, so $y \in \varnothing$. This is false, so $\bigcap_{X \in \mathcal{P}(E)} X$ must be the empty set.

(ii) To show that $E = \bigcup \mathcal{P}(E)$, we need to show that $E \subset \bigcup \mathcal{P}(E)$ and $\bigcup \mathcal{P}(E) \subset E$.

If $x \in E$, then $\{x\} \subset E$ and so $\{x\} \in \mathcal{P}(E)$.

Thus $x \in \bigcup \mathcal{P}(E)$.

This means that $\bigcup \mathcal{P}(E) \subset E$.

Also, if $x \in \bigcup \mathcal{P}(E)$, then for some $A \in \mathcal{P}(E)$, $x \in A$. That is, $x \in A$ and $A \subset E$, so $x \in E$.

This means that $\bigcup \mathcal{P}(E) \subset E$.

Our proof is now complete.

(iii) Let $x \in E$. Then $x \subset \bigcup E$ (see Example 2, Section 1.4.1) and so $x \in \mathcal{P}(\bigcup E)$.

Solution to SAQ 9

(i) The empty set, $\varnothing$.

(ii) The set A itself.

(iii) $\{\varnothing, \{\varnothing\}\}$.

(iv) $A \cup \{\varnothing\}$.

Solution to SAQ 10

(i) This is *not* a valid application: x has not been restricted to be a member of a known set. (See **Halmos**: *page 11*.)

(ii) This *is* a valid application. The set is the intersection of A and B.

Solution to SAQ 11

(i) This statement is true only if $\{\varnothing\} \subset A$, which is the same as $\varnothing \in A$. This is not true for all sets A; for example, $\varnothing \notin \varnothing$.

(ii) This statement is never true.

We can see this as follows.

If A is a finite set with n elements, then $\mathcal{P}(A)$ has 2^n elements (see **Halmos**: *page 20, line 8*) and hence $\mathcal{P}(A) = A$ cannot be true.

We now give a proof for the general case.

Let us assume that $\mathcal{P}(A) = A$.

Then $A \in \mathcal{P}(A)$ so $A \in A$.

Consider the set

$$X = \{x \in A : x \notin x\}.$$

Then $X \subset A$ so $X \in \mathcal{P}(A)$; hence, by our assumption, $X \in A$.

Consider now whether $X \in X$.

If $X \in X$, then, by the definition of X, $X \notin X$; if $X \notin X$, then, since $X \in A$, $X \in X$, so we have a contradiction in each case.

Hence $\mathcal{P}(A) \neq A$.

(iii) This statement is always true.

Solution to SAQ 12

If $x, y \in A$ (x and y distinct), then $\{x\}$ and $\{y\}$ are proper subsets. Thus a set with no proper subset must be either empty or a singleton.

The Axiom of Powers asserts that the collection of all subsets of E does form a set. The Axiom of Specification then asserts that the collection of all singleton or empty subsets of E is indeed a set, as follows.

$$I = \{A \in \mathcal{P}(E) : x \in A \text{ and } y \in A \text{ implies } x = y\}.$$

(Note that $\varnothing \in I$, for, since $\varnothing$ has no elements,

$$x \in \varnothing \text{ and } y \in \varnothing \text{ implies } x = y$$

is vacuously true, i.e., not false.)

Unit 2   Set Constructions

"[When followed in the proper spirit], there is no study in the world which brings into more harmonious action all the faculties of the mind than the one [mathematics] of which I stand here as the humble representative and advocate. There is none other which prepares so many agreeable surprises for its followers, more wonderful than the transformation scene of a pantomime, or, like this, seems to raise them, by successive steps of initiation to higher and higher states of conscious intellectual being."

J. J. Sylvester
A Plea for the Mathematician, *Nature*, Vol. 1
p. 261.

Contents

Set Books

P. R. Halmos, *Naive Set Theory*, paperback edition 1972 (Van Nostrand Reinhold).
I. N. Herstein, *Topics in Algebra*, paperback edition (Xerox College/T.A.B.S, 1964).
B. Mendelson, *Introduction to Topology*, paperback edition 1972 (Allyn and Bacon).
M. L. Minsky, *Computation: Finite and Infinite Machines*, paperback edition 1972 (Prentice-Hall).

It is essential to have these books; the course is based on them and will not make sense without them.

Unit 2 is based on **Halmos**, Sections 6–10.

Conventions

Before working through this correspondence text make sure you have read *A Guide to the Course: Topics in Pure Mathematics.*

References to the Open University Mathematics Foundation Course Units (The Open University Press, 1971) take the form *Unit M100 3, Operations and Morphisms.*

2.0 INTRODUCTION

The purpose of this unit is to revise, consolidate, and in some regions extend, a number of the basic ideas introduced in the Foundation Course. The material covers Sections 6 to 10 of **Halmos**. We shall illustrate it with occasional examples from **Herstein** and **Mendelson**, and we shall make a special note of any instances of notational or terminological differences between **Halmos** and the Foundation Course. These differences are irritating, but they are inescapable facts of mathematical life and you must accustom yourself to living with them.

Exercises occur throughout **Halmos**, either explicitly under the heading "Exercise" or embedded in the text where their presence is indicated as, for example, on *page 38*, *lines 6 and 7*, by the words "(proof?)" or "(example?)". Many of these exercises appear in this commentary either as worked examples or, later, as self-assessment or other questions, possibly in a slightly amended form. If you wish, you may, of course, attempt them as they arise in **Halmos**.

Aims

After working through this text, you should be able to:

(i)   manipulate ordered pairs;
(ii)   recognize if a relation is reflexive, symmetric or transitive;
(iii) verify simple properties of functions, their inverse images, and composite functions;
(iv)   understand how fundamental mathematical ideas such as *function* and *relation* can be placed in a purely set-theoretic context;
(v)   describe the Cartesian product of a family of sets as a set of functions.

2.1 ORDERED PAIRS

2.1.0 Introduction

"The concept of an ordered pair could have been introduced as an additional primitive, axiomatically endowed with just the right properties, no more and no less." This is a quotation from **Halmos**, *page 25*, and it describes fairly accurately the treatment of ordered pairs in the Foundation Course. (*Primitive* means *undefined concept*.) The alternative is to specify the concept of ordered pair in terms of set properties, and this is our approach here. The Cartesian product of two sets is then a set of ordered pairs.

2.1.1 Ordered Pairs as Sets

READ **Halmos** : *page 22, line 1, to page 23, line −5*.

The embedded exercise is discussed in Note (ii) below.

Notes

(i) **Halmos** : *page 22, lines −9 to −6*.
  We could equally well take

$$\mathcal{C} = \{\{d, b, c\}, \{c, b\}, \{c\}, \{b, d, a, c\}\}$$

or

$$\mathcal{C} = \{\{b, c\}, \{a, c, d, b\}, \{c\}, \{d, c, b\}\},$$

for these are the same set and give the same order as that in the text.

Of course, we *do* know what *order* means: if not, would we be able to read? However, we have not yet stated what it is to mean mathematically.

(ii) **Halmos** : *page 23, lines 8 to 12*.
  For the purpose of this exercise, take "order" to mean the particular kind of order classified in the Foundation Course as "total order": that is, if $x \in A$ and $y \in A$ and x and y are not the same element, then *either* x "comes before" y *or* y "comes before" x. One of the difficulties of the exercise arises from the fact that we may use only those ideas so far defined in **Halmos**. Concepts such as numbers or one-to-one correspondence are beyond us at present.

Solution to the Exercise

In order that a collection $\mathcal{C}$ of subsets of A should correspond to an ordering of the kind described, it is necessary that we should be able to carry out, with those subsets, the procedure described in **Halmos**. We are now asked to describe the properties which $\mathcal{C}$ must possess in order to make this possible.

First, we see that the procedure fails with such a collection as

$$\{\{a\}, \{a, b\}, \{a, c\}, \{a, b, c\}\}$$

and we observe that one essential property is that every subset must *either* contain, *or* be contained in, every other subset. One way of formalizing this requirement is to stipulate:

$$\text{for every } Y \in \mathcal{C}, \{X \in \mathcal{C} : X \subset Y\} \cup \{X \in \mathcal{C} : Y \subset X\} = \mathcal{C}. \tag{1}$$

Now application of the procedure to a collection $\mathcal{C}$ that satisfies (1) alone will give us a "chain" of subsets. For example, if

$$A = \{a, b, c, d, e, f\}$$

and

$$\mathcal{C} = \{\{c, d\}, \{c, d, e, a\}, \{c, d, e, a, b\}\},$$

then (1) is satisfied and the "chain" is

$$\{c, d\} \subset \{c, d, e, a\} \subset \{c, d, e, a, b\}.$$

But there are three things wrong in this example:

(a) the "chain" does not start with a singleton;
(b) the "chain" does not end with the whole set A;
(c) the "chain" does not increase by steps of single elements.

We see that (a) and (b) are easily corrected: we need the stipulation

A itself is an element of $\mathcal{C}$, and one element of $\mathcal{C}$ is a singleton.   (2)

(1) and (2) together ensure that we have a "chain" which "starts" at the right place (some singleton) and "ends" at the right place (A itself).

To deal with difficulty (c), we have to impose the condition that each step gives us a single element. One method is to use the third stipulation:

for every $Y \in \mathcal{C}, (Y \neq A)$, there is a Y' in $\mathcal{C}$ such that $Y' - Y$ is a singleton.   (3)

Are (1), (2) and (3) necessary and sufficient conditions for the completion of Halmos' procedure? If you are not sure, take heart: you have at least learnt that the formalization of an apparently simple intuitive idea may be a matter of some difficulty! We shall have considerably more to say about order later.

(iii) *Halmos*: *page 23, lines -16, -15.*
Note that we are told here that

$$(a, a) = \{\{a\}\}, not \text{ that } (a, a) = \{a\}.$$

(Both $\{\{a\}\}$ and $\{a\}$ are singletons, but the unique element of the first is the singleton of a while the unique element of the second is a itself.)

(iv) *Halmos*: *page 23, line -12.*
"... it follows that ..." actually involves two steps, first that $\{a\} = \{x\}$ and then that $a = x$.

SAQ 1

In order to appreciate the definition of an ordered pair as a set, specify each of the following as a set:

(i) $(a, b) \cup (c, d)$

(ii) $(a, b) \cup (a, b)$

(iii) $(a, b) \cap (a, b)$

(iv) $(a, b) - (b, a)$, where $b \neq a$

(v) $(a, b) - (a, a)$, where $b \neq a$.

(Solution is given on p. 34.)

2.1.2   The Cartesian Product of Two Sets

To specify a set, we use the Axiom of Specification. This requires us to quote an already known set A, some of whose elements are to make up the required set, together with a sentence $S(x)$ describing those elements. Now if we have two properly defined sets X and Y, and wish to define their Cartesian product, we must first define a set of which this is to be a subset. The first part of the next reading passage shows that the required set is

$$\mathcal{P}(\mathcal{P}(X \cup Y)).$$

The next part reverses the idea: can we express any set of ordered pairs as the Cartesian product of two suitable sets?

*READ **Halmos**: page 23, line -4 to page 24, line -6.*

Notes

(i)   ***Halmos**: page 24, lines 1 to 3.*
      Note that,

   if $P \subset Q$, then $P \in \mathcal{P}(Q)$;

similarly,

   if $X \subset \mathcal{P}(Y)$, then $X \in \mathcal{P}(\mathcal{P}(Y))$.

You may find that reasoning of this kind, or in considerably more complicated instances and proofs, is better appreciated if you construct a diagrammatic layout. Notice that the value of such a procedure lies almost wholly in constructing *your own* diagram. We give two examples (one below and one at the end of Note (ii)) merely as specimens.

Conventions

We shall write:

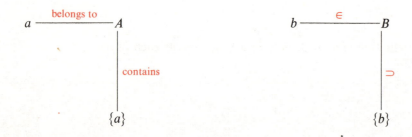

*Argument 1, **Halmos**: page 24, lines 1 to 3.*
In this case, we have:

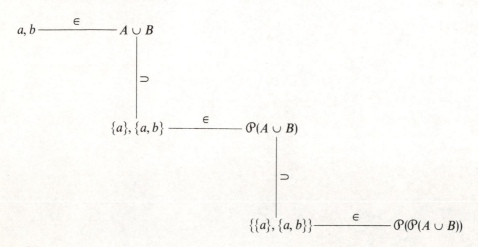

For practice with power sets, take $A = \{a\}$ and $B = \{b\}$ and write out all the elements of $\mathcal{P}(\mathcal{P}(A \cup B))$. You should find that

$A \cup B$ has two elements,

$\mathcal{P}(A \cup B)$ has four elements,

and $\mathcal{P}(\mathcal{P}(A \cup B))$ has sixteen elements.

(If you have listed only fifteen, you have probably overlooked the fact that $\varnothing$ and $\{\varnothing\}$ are both elements of $\mathcal{P}(\mathcal{P}(A \cup B))$.)

(ii) **Halmos**: *page 24, lines 15 to −6.*
We apply the reasoning in the text to an example.

Example

Let R, the set of ordered pairs, be

$$\{(a, b), (a, c), (d, c)\}.$$

Then, by the definition of ordered pairs,

$$R = \{\{\{a\}, \{a, b\}\}, \{\{a\}, \{a, c\}\}, \{\{d\}, \{d, c\}\}\}$$

so

$$\bigcup R = \{\{a\}, \{d\}, \{a, b\}, \{a, c\}, \{d, c\}\}.$$

(Remember that $\bigcup R$ is an abbreviation for $\bigcup_{X \in R} X$.)

Taking the argument one step further, we have

$$\bigcup\bigcup R = \{a, b, c, d\} = S, \text{ say.}$$

Then, indeed, $R \subset S \times S$.

A diagrammatic form of this argument, using similar conventions to those in the previous diagram, is

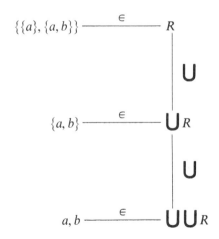

We see that $S \times S$ has sixteen elements. If this is too large for convenience, we take A and B as the projections of R on to the first and second co-ordinates, that is, the set of all first co-ordinates and the set of all second co-ordinates, respectively. We thus obtain

$$A = \{a, d\}, \qquad B = \{b, c\}$$

and, again,

$$R \subset A \times B.$$

SAQ 2

(i) Give an example to show that the set $A \times B$ is not necessarily equal to $B \times A$.

(ii) Is it necessarily true that

$$A \times (B \times C) = (A \times B) \times C?$$

(Note that when we generalize the Cartesian product to more than two sets in Section 2.4.2, we shall find that $A \times B \times C$ is usually neither of the above sets.)

(iii) Show by an example that not every set of ordered pairs is equal to the Cartesian product of two sets.

(Solution is given on p. 34.)

2.1.3 Summary

READ Halmos: page 24, line −5 to the end of the section, page 25.

The exercise appears as SAQ 3.

SAQ 3

Work the whole of the exercise at the end of the section, *Halmos: page 25*.

(Solution is given on p. 34.)

2.2 RELATIONS

2.2.0 Introduction

There is little in this section which will be new to you if you have studied the Foundation Course. In Section 2.2.1 a *relation* is defined as a set of ordered pairs. In Section 2.2.2 the *domain* and *range* of a relation are introduced, and some of the possible properties of relations in sets are considered. In Section 2.2.3 the correspondence between *partition* and *equivalence relation* is demonstrated.

2.2.1 Relations as Sets

READ **Halmos**: *page 26, line 1 to page 27, line 16.*

Note

Halmos: *page 26, lines − 2 and − 1.*
Note that any set of ordered pairs *is* a relation, but do not be misled into thinking that, given any set of ordered pairs, it is proper to ask whether or not the relation is reflexive. Such a property belongs to a relation *in a set*, not to the relation alone.

SAQ 4

If $X = \{a, b\}$ and

$$R = \{(\varnothing, \varnothing), (\varnothing, \{a\}), (\varnothing, \{b\}), (\varnothing, \{a, b\}),$$
$$(\{a\}, \{a\}), (\{a\}, \{a, b\}),$$
$$(\{b\}, \{b\}), (\{b\}, \{a, b\}),$$
$$(\{a, b\}, \{a, b\})\},$$

describe in words the relation R.

(HINT: It is not helpful here to return to the definition of ordered pairs as sets.)

(Solution is given on p. 35.)

2.2.2 Some Properties of Relations

*READ **Halmos**: page 27 from the third paragraph ("In the preceding section . . .") to page 28, line 3.*

The embedded exercise appears as SAQ 6 in Section 2.2.3.

Notes

(i) **Halmos**: *page 27, lines* -20 *to* -15.
The word *domain*, here used by Halmos in the context of relations, conforms to his later use of the same word in the context of functions; it is used in the same sense, for functions, in the Foundation Course and by most authors. The word *range*, however, is a different matter. Halmos uses *range* (of a function) to mean the set which, in the Foundation Course, we referred to as the *image set*, whereas some other authors use *range* to mean the set which, in the Foundation Course, we referred to as the *codomain*.

(ii) **Halmos**: *page 27, lines* -4 *to* -2.
We emphasize our remark in Section 2.2.1. It is common practice to say, for instance, "this relation is reflexive" but what is meant is always "this relation is reflexive in this set", where both the relation *and* the set are specified or understood.

Some students have difficulty with the properties of relations in sets. Halmos redefines these properties later (**Halmos**: *page 41*) in a form which may prove helpful. The following diagrammatic summary may be useful.

We shall represent $x\,R\,y$ by $\overset{\displaystyle\frown}{\underset{x\qquad\; y}{\bullet\qquad\bullet}}$

(The dots in the diagrams represent elements, but these elements are not necessarily distinct.)

Reflexivity For *every* element $x \in S$

Symmetry For *every* element $x \in S$,
for *every* element $y \in S$,

implies

Transitivity For *every* element $x \in S$,
for *every* element $y \in S$,
for *every* element $z \in S$,

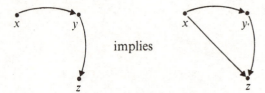

implies

Included in this is the special case when $x = z$.

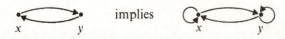

implies

SAQ 5

Consider the relation

$$R = \{(p, p), (q, q), (p, q), (q, p)\}.$$

Is R

(a) reflexive

(b) symmetric

(c) transitive

in the set (i) $X = \{p, q, r\}$?
            (ii) $Y = \{p, q\}$?

(Solution is given on p. 35.)

2.2.3 Equivalence Relations and Partitions

READ **Halmos** : page 28, line 4 to the end of the section, page 29.

The embedded exercise appears as SAQ 7.

Notes

(i) **Halmos** : page 28, lines 12 to 14.
     The notation X/R for the set of equivalence classes is virtually universal; it is sometimes read as "the quotient set of X by R". (The reason for the word "quotient", and hence for the fraction bar, will become evident later.)

     On the other hand, the notation for "the equivalence class to which x belongs" is very much a matter of the author's choice; Halmos uses x/R, the Foundation Course used $[x]$, and other authors use other symbols.

(ii) **Halmos** : pages 28 and 29, the last two paragraphs of the section.
      This proof is effectively that given in Unit M100 19, Relations.

SAQ 6

Work the exercise, **Halmos** : page 27, lines -2 and -1, using the same set, $X = \{a, b, c\}$, in each case.

(Solution is given on p. 35.)

SAQ 7

Work the exercise, **Halmos** : page 28, lines 15 to 17.

Note that your condition will specify a set (not necessarily a partition) even if R is not an equivalence relation.

(Solution is given on p. 36.)

SAQ 8

Herstein : page 9, Problem 10.

(Herstein uses $a \sim b$ to mean $a \, R \, b$. Also, Herstein is using objects like the real numbers and symbols like $<$ which Halmos has not yet introduced.)

(Solution is given on p. 36.)

SAQ 9

Herstein : page 9, Problem 10.

(Property 1 of an equivalence relation is $a \sim a$ for all $a \in A$.)

(Solution is given on p. 37.)

2.3 FUNCTIONS

2.3.0 Introduction

In Section 8, Halmos defines a *function* to be a special kind of relation, and then discusses some special functions which occur frequently in mathematics.

If you have studied the Foundation Course, you will find little new material in Section 2.3.1 apart from some minor differences of terminology. In Section 2.3.2, however, there appear some special functions which we shall use in many different areas of this course later on.

2.3.1 Notation and Terminology

READ Halmos: page 30, line 1 to the middle of page 31 (i.e., to "...writing $f(X) = Y$.").

Notes

(i) *Halmos: page 30, lines 10 to 12.*
The words *map, mapping, transformation, correspondence, operator* and *function* are not synonymous for all authors. The only term among these which has the same meaning for all (modern) authors is *function*. For instance, in the Foundation Course, *mapping* was a more general term than *function* (we talked, for example, of one-many mappings, and these are not functions) and *operator* was used only for special functions whose domains were sets of functions. We shall see more differences of this kind as we meet new authors.

(ii) *Halmos: page 30, lines −11 and −10.*
The notation Y^X for the set of all functions from X to Y may look peculiar, but it is very reasonable, as we shall see in SAQ 10. It is also very useful and very common.

(iii) *Halmos: page 31, lines 4 and 5.*
Remember that Halmos uses *range* to mean what the Foundation Course called "the image set".

(iv) *Halmos: page 31, lines 8 to 15.*
An example of the unlikely situation referred to is as follows.

Example

Suppose

$$A = \{a, b\}, X = \{a, b, A\}, Y = \{x, y, z\}$$

and the function $f: X \longrightarrow Y$ is defined by

$$f(a) = x, \; f(b) = y, \; f(A) = z.$$

Then $f(A)$ is ambiguous: it could mean either the image of the element A, which would be z, or the image of the subset $\{a, b\}$, which would be $\{x, y\}$.

As Halmos says, the notation is bad but not catastrophic; we can rely on the context or, where necessary, on further verbal stipulations to avoid this confusion.

(v) Since we are discussing the images of subsets of domains, we take the opportunity to describe a different approach which will become important in its own right in all parts of this course later on. To show it here we anticipate a little, and assume that we have two functions f and g, where f maps A to B and g maps B to C. We shall for the present write $g \circ f$ for the composite map from A to C.

We have the (triangular) commutative diagram:

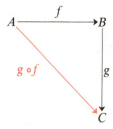

Now we regard these three maps as mapping the *elements* of their domains to their images; we refuse to allow them to map *subsets* of their domains. Remember that, if a is an element of A, then $\{a\}$ is a subset of A. If we wish to map subsets, we define a new function f_* such that, if X is a subset of A, then

$$f_*(X) = \{y \in B : y = f(x), x \in X\},$$

so that f_* maps $\mathcal{P}(A)$ to $\mathcal{P}(B)$. (Note that $f_*(\varnothing) = \varnothing$.) We now have three sets, $\mathcal{P}(A)$, $\mathcal{P}(B)$ and $\mathcal{P}(C)$, and hence new functions f_*, g_* and $(g \circ f)_*$, giving the new commutative diagram:

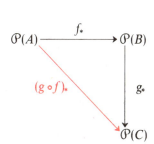

where $(g \circ f)_* = g_* \circ f_*$.

We use the starred maps to map subsets and the unstarred maps to map elements, and all is well. This may well seem a major fuss about a minor triviality, and so it would be if it ended there, but, as we shall see, the general process by which one commutative diagram can be made to produce another (of which this must be about the simplest possible example) is a very powerful tool in both algebra and topology. You will remember that *morphism* was a central theme in the Foundation Course and that a morphism can be represented by a commutative diagram. Here we are beginning to think about a morphism of diagrams.

SAQ 10

If

$$X = \{a, b\} \text{ and } Y = \{c, d, e\},$$

how many elements (i.e. functions) are there in the sets Y^X and X^Y?

(We are not yet entitled to ask questions beginning "How many . . . ," but answer the question anyway. Its purpose is to give some justification for the Y^X notation.)

(Solution is given on p. 38.)

2.3.2   Some Special Functions

Throughout this course we shall make considerable use of the simple special functions defined in the next reading passage; it is worth memorizing their names.

*READ **Halmos**: page 31, line 20 to page 32, line 18.*

Notes

(i)   ***Halmos**: page 31, line 21.*
The terms *embedding* and *injection* are used by other authors with different meanings. Stick to the phrase *inclusion map*.

(ii) ***Halmos**: page 31, line −8.*
The idea of the *identity map* may seem very trivial, but, as we mentioned in Section 2.3.1, we shall often be concerned with the construction of commutative diagrams. Now it quite often happens that a commutative diagram has not enough "edges" for what we want to do. The insertion of the appropriate identity map may well be a vital step.

Structurally, of course, identity maps are essential. As one very minor example, the special importance of the real numbers 0 and 1 rests on the fact that each of the functions

$$x \longmapsto x + 0 \text{ and } x \longmapsto 0 + x$$

and

$$x \longmapsto x \times 1 \text{ and } x \longmapsto 1 \times x$$

is the identity map on the set of real numbers.

(iii) ***Halmos**: page 31, line −3 to page 32, line 5.*
Let $A \subset X$, $f : A \longrightarrow Y$ and $F : X \longrightarrow Y$.

The F is an *extension* of f if and only if the following diagram is commutative, where $i : A \longrightarrow X$ is the *inclusion mapping*.

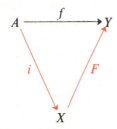

(iv) ***Halmos**: page 32, line 18.*
The *canonical map* from X to X/R is often called the *natural map*, as in the Foundation Course.

2.3.3   Functions and Equivalence Relations: Characteristic Functions

*READ **Halmos**: page 32, line 19, to the end of the section on page 33.*

The embedded exercise and the final exercise appear as SAQ 11 and SAQ 12 respectively.

Notes

(i)   ***Halmos**: page 32, lines −15 to −12.*
It is interesting to note that the basic severity of Halmos' approach has caused him to use quite a sophisticated example as his introduction to one-to-one correspondence. Later (our Section 2.5), using his definition of inverse image, he can define a function f from X to Y to be one-to-one if and only if the inverse image of each singleton in the range of f is a singleton in X. The function f will be one-to-one and onto if the inverse image of each singleton in Y is a singleton in X.

(ii)   ***Halmos**: page 32, line −6.*
"Natural numbers" is another of *those* phrases. You have to check what the author means. Some authors do not include zero in the set of natural numbers.

(iii)   ***Halmos**: page 32, last line.*
Many students would probably accept, with fewer qualms, definitions like

> 1 is the equivalence class of all those sets which can be put in one-to-one correspondence with $\{\varnothing\}$

because "can be put in one-to-one correspondence with" is an equivalence relation in a set of sets. The difficulty here is that the equivalence class referred to above is necessarily a set (in spite of the presence by tradition of the word "class") and hence must satisfy the Axiom of Specification. Now this requires us to specify a properly defined set A of which our equivalence class is a subset— and where shall we find such a set A? We shall see more of Halmos' approach to this problem in *Unit 3, Sets and Numbers*.

(iv)   ***Halmos**: page 33, lines 9 and 10.*
As we shall see later (e.g., ***Halmos**: page 52, line 15*), the word "between" is very important. We say that there is a one-to-one correspondence *between* two sets if there exists a function from one to the other which is one-to-one *and onto*.

(v) *Halmos*: *page 33, final paragraph.*

If *A*, *B* and *C* are subsets of a set *E*, the following diagram shows the values of $\chi_A(x)$, $\chi_B(x)$ and $\chi_C(x)$ for any element *x*, in each of the eight regions.

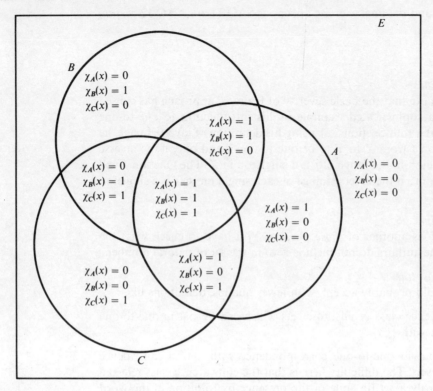

SAQ 11

Work the exercise in parentheses, *Halmos*: *page 32, line* − *10.*

(Assume that the Cartesian product is non-empty.)

(Solution is given on p. 38.)

SAQ 12

Work the exercise, *Halmos*: *page 33*, at the end of the section.

(It is worth noting that this is an example of a situation where intuition, which is so often helpful, now leads most people astray and a formal approach is virtually essential.)

(Solution is given on p. 38.)

SAQ 13

Mendelson: *page 15, Exercise 7.*

(Here, again, we are asking you to perform arithmetical calculations which have not yet been defined in *Halmos*.)

(Solution is given on p. 39.)

2.4 FAMILIES

2.4.0 Introduction

At last we meet a substantial new idea. It is not at first sight a particularly important one, but it is used implicitly throughout mathematics. At this basic stage, the major use of the idea of a *family* of sets is in generalizing properties of operations like unions which have previously been established only for special cases. Later on, in topology, the idea becomes something rather more than the notational matter which it is here.

One important use of families is the construction of a larger set incorporating each member of the family in one big set. This is the generalized Cartesian product.

2.4.1   Notation and Terminology: Unions and Intersections of Families

The next reading passage expresses very simple ideas in the notation of families. You should treat the passage as an opportunity to learn to read the symbols.

READ **Halmos**: *page 34, line 1 to page 35, line −5.*

(The embedded exercise, *page 35, line 10*, may be ignored.)

Notes

(i) **Halmos**: *page 34, lines 1 to 21.*
   "There are occasions when the range of a function is deemed to be more important than the function itself." This is indeed the case here.

   We can summarize *lines 5–8* as follows:

$$x \ : I \longrightarrow X$$
   family index   indexed
         set   set

$$x \ : i \ \longmapsto \ x_i$$
   family index   term of
                   the family

   Notice, from *line 7*, that the family *is* the function from I to X, but that, on *lines 15 and 16*, where X has been replaced by $\mathcal{P}(X)$, we have a "family of subsets". Two lines further on we have "the union of the family", by which is meant the union of the subsets which form the range of the function. This latter use is in fact the way in which the family terminology is normally used.

(ii) **Halmos**: *page 35, lines 5 to 9.*
   Any difficulty you have here is purely notational. We translate the notation in the simplest case in the following example. All that is being said is that, to form the union of a family, you can form unions of parts of the family and then union the lot, provided that every index is covered at least once. More explicitly, to form the union of a family $\{A_k\}$, indexed by K, we may first write K as a union of subsets indexed by J say:

$$K = \bigcup_{j \in J} I_j,$$

then, for each j, form the union of the family indexed by I_j:

$$\bigcup_{i \in I_j} A_i,$$

and, finally, take the union of all sets of the above form:

$$\bigcup_{j \in J} \left(\bigcup_{i \in I_j} A_i \right).$$

19

Example

Let

$$K = \{a, b, c\}$$

and

$$\{A_a, A_b, A_c\}$$

be some family of sets indexed by K.

We shall now write K as a union of subsets,

$$K = I_r \cup I_s \text{ say, where}$$

$$I_r = \{a, b\} \text{ and } I_s = \{c\}.$$

Writing

$$J = \{r, s\},$$

we see that

$$\{I_r, I_s\}$$

is a family of sets indexed by J.

Now the left-hand side of **Halmos**: *line 8 is*

$$\bigcup_{k \in K} A_k = A_a \cup A_b \cup A_c$$

and the right-hand side is

$$\bigcup_{j \in J} \left(\bigcup_{i \in I_j} A_i \right) = \left(\bigcup_{i \in I_r} A_i \right) \cup \left(\bigcup_{i \in I_s} A_i \right)$$
$$= (A_a \cup A_b) \cup A_c.$$

So we have, merely, that

$$A_a \cup A_b \cup A_c = (A_a \cup A_b) \cup A_c.$$

(iii) **Halmos**: *page 35, lines 11 and 12.*

In the phrase "an empty union" the word "empty" may be taken as referring to the domain of the family, but, since the union of a family indexed by the empty set is itself empty, it does no harm to take the word as referring to the union too. On the other hand, (as explained in **Halmos**: *page 19*), the intersection of a family of subsets of E indexed by the empty set is E itself, so that if, in the phrase "an empty intersection", we take "empty" as referring to the domain of the family, we have "an empty intersection is not necessarily empty". Because of this apparent contradiction, the phrase "an empty intersection" is best avoided.

2.4.2 Cartesian Products of Families

This section is based on the material in **Halmos**: *page 36, line 3 to page 37, line 3*, but it is not essential for you to read this passage as its content is covered in our commentary.

The Cartesian product of two sets appears nearly everywhere in mathematics, since relations and functions both involve the idea implicitly. It is also necessary to extend the idea to Cartesian products of many sets, and the simplest and usual way of doing this is by using the notation of families. Of course, there would be no difficulty in defining a Cartesian product of three sets by first defining an *ordered triple* and then defining $A \times B \times C$ as the set of all ordered triples of the form (a, b, c) where $a \in A$, $b \in B$ and $c \in C$, and similarly for any finite number of sets. This procedure becomes far from satisfactory, however, if we are faced with the problem of forming a Cartesian product of infinitely many sets. It is this difficulty that prompts us to seek a different approach.

Let us look again at the set

$$A \times B = \{(a, b) : a \in A, b \in B\}.$$

Now in **Halmos**: *page 24*, we saw that, if $R \subset A \times B$, then it is natural to look at the *projection* of R on to the first and second co-ordinates. We have

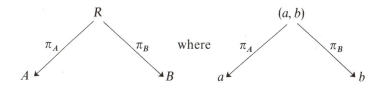

It is precisely these projections which make the Cartesian product useful. For suppose that you are told that the first co-ordinate of $r \in R$ is p, and the second is q; then you can immediately reconstruct the element $r \in R$ as $r = (p, q)$.

This all seems pretty trivial, but in fact it has far-reaching consequences. In order to describe a *member* of $A \times B$, all we really need to know are the images under its projections in the component sets A and B. These projections can be represented by tags attached to the elements of $A \times B$. The projection on to A might be represented by a red tag, and the projection on to B might be represented by a white tag. For each element, its image under the projection π_A is written on the red tag, and its image under the projection π_B is written on the white tag. Thus, for

$$A = \{a_0, a_1\} \text{ and } B = \{b_0, b_1\},$$

we obtain the following representation of $A \times B$:

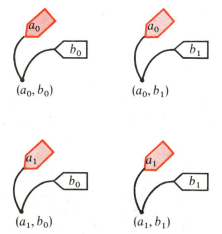

The tags can be conveniently formalized as functions. To each of the members (a, b) of $A \times B$, there corresponds a function f

$$f: \{\text{red, white}\} \longrightarrow A \cup B,$$

where

$$f(\text{red}) = \text{contents of red tag}$$
$$= \text{projection of } (a, b) \text{ on } A = a,$$
$$f(\text{white}) = \text{contents of white tag}$$
$$= \text{projection of } (a, b) \text{ on } B = b.$$

The result of all this is that we can think of $A \times B$ in two distinct ways:

(i) as $\{(a, b): a \in A, b \in B\}$;

(ii) as $\{f: \{\text{red, white}\} \longrightarrow A \cup B, \text{ where } f(\text{red}) \in A \text{ and } f(\text{white}) \in B\}$.

Now (i) and (ii) are distinct sets—one is a set of ordered pairs, the other is a set of functions. But both have the properties we wish the Cartesian product to have. That is, they have projections on to the components A, B. (They also have much more, but we shall examine this after we have defined the Cartesian product of a family.)

Suppose we now have a family, $\{A_i: i \in I\}$, of subsets of some set S. We wish to define a Cartesian product of these subsets. The easiest approach is the functional approach of (ii) above. Instead of objects with red and white tags attached to them, we now speak of the ith tag, where i is some member of the index set I. The contents of the ith tag is to be an element of A_i. The set of functions we want is then

$$Z = \{f: I \longrightarrow \bigcup_{i \in I} A_i, \text{ where } \forall i \in I, f(i) \in A_i\}.$$

Each element of Z is a function f. Each function associates with the element $i \in I$ an element $f(i) \in A_i$.

SAQ 14

To be sure that Z is a set, the Axiom of Specification requires a previously known set of which Z is a subset. Provide such a set.

(Solution is given on p. 40.)

The set Z, the Cartesian product of the family $\{A_i\}$, should come equipped with projections on to the components A_i. These are given by:

$$\pi_i: Z \longrightarrow A_i, \text{ where } \pi_i(f) = f(i) \qquad (f \in Z).$$

Thus the ith projection of f is just the image of i under f (the contents of the ith tag).

Let us now relate this to the initial idea of ordered pairs. Think of the index set I in two forms. First as dots scattered about, and, second, as dots strung out on a line. Take $I = \{R, O, Y, G, B, V\}$.

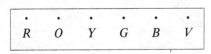

Now *each* $f \in Z$ maps I to $\bigcup_{i \in I} A_i$. We shall abuse the notation by simply placing the image of i under f at the dot representing i. Thus we have

Either of these diagrams can be taken to represent the element f in Z. By placing brackets and commas in the second, and suppressing dots, we get the usual notation for "ordered n-tuples", where the order has been provided by some order imposed on the index set. Thus

$$(f(R), f(O), f(Y), f(G), f(B), f(V))$$

is an "ordered 6-tuple".

It is not the order which is important, but the ability to pick out or project on to the various components.

Convention

We denote by

$$\underset{i \in I}{\mathsf{X}} A_i$$

either the set of functions Z *or* the set of n-tuples, whichever is convenient. Usually, if I is a small set, we think of n-tuples. If I is large, we think of functions.

The Universal Property of $\underset{i \in I}{\mathsf{X}} A_i$

We shall now indicate why the Cartesian product is so useful. We shall need to use the idea of composite functions which Halmos does not introduce until the next section.

Let $\{A_i : i \in I\}$ be a family of subsets of S. Let Z be the set of functions which represent $\underset{i \in I}{\mathsf{X}} A_i$. Suppose we are confronted with a new set W such that, for each $i \in I$, there is a function

$$g_i : W \longrightarrow A_i.$$

These functions g_i look a bit like projections from W to the A_i, and so we might expect W to be related to the Cartesian product Z in some way. This is indeed the case.

We have, $\forall i \in I$,

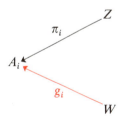

The property that makes Z useful and interesting is that there is a *unique* function

$$h : W \longrightarrow Z$$

which, $\forall i \in I$, makes the diagram commutative.

In order to study W and all its functions g_i, it is perfectly adequate to study the function h, and the projections π_i, which are easy to deal with. Thus h codes or stores up all the information provided by the functions g_i. This is why the Cartesian product is a useful construction.

To see how h arises, we choose an element $w \in W$. We want to specify $h(w) \in Z$, and $\forall i \in I$ we want

$$\pi_i(h(w)) = g_i(w).$$

Now $h(w) \in Z$ means $h(w)$ is a function:

$$h(w) : I \longrightarrow \bigcup_{i \in I} A_i.$$

We now require the following commutative diagram:

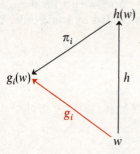

The obvious choice is to specify

$$h(w): i \longmapsto g_i(w) \qquad (i \in I)$$

for then

$$\pi_i(h(w)) = (h(w))(i) = g_i(w),$$

as required.

Remarks

(i) All this may seem to be mindless diagram chasing. Diagram chasing it is: mindless it is not. It is very simple, but very important.

(ii) To specify h we *chose* to make

$$h(w): i \longmapsto g_i(w) \qquad (i \in I).$$

It seems entirely reasonable to make this choice, but there are difficulties. Indeed, to be sure that Z is not empty, we must add an additional set axiom! We shall discuss this in later units.

(iii) The uniqueness of h is a matter of chasing diagrams and/or symbol manipulation, as follows:

Suppose k also does the job. For each $i \in I$, we have

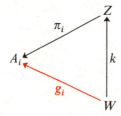

where

$$\pi_i \circ k = g_i.$$

To show that k and h are equal, we must show that, $\forall w \in W$, $h(w) = k(w)$. For any $w \in W$, let us examine $k(w)$. Now $k(w) \in Z$ and so it is a function

$$k(w): I \longrightarrow \bigcup_{i \in I} A_i.$$

To show that $k(w) = h(w)$, we must show that, $\forall i \in I$,

$$(h(w))(i) = (k(w))(i).$$

Now

$$(h(w))(i) = \pi_i(h(w))$$

and

$$(k(w))(i) = \pi_i(k(w)).$$

But

$$\pi_i \circ h(w) = g_i(w) = \pi_i \circ k(w),$$

so $h(w)$ and $k(w)$ are equal functions, and therefore h and k are the same function!

SAQ 15

Suppose that I is an index set and $\{A_i : i \in I\}$ is an indexed family of subsets of S. Suppose that W and V respectively have the following maps associated with them:

$$g_i : W \longrightarrow A_i \qquad (i \in I),$$

$$h_i : V \longrightarrow A_i \qquad (i \in I),$$

where both W and V share the universal property possessed by $\displaystyle \bigtimes_{i \in I} A_i$. Thus the elements of the set $\{g_i : i \in I\}$ are playing the roles of the projections for W, and the elements of the set $\{h_i : i \in I\}$ are playing the roles of the projections for V.

Show that there exist two functions

$$\sigma : W \longrightarrow V \quad \text{and} \quad \psi : V \longrightarrow W,$$

such that

$$\sigma \circ \psi = \text{identity on } V$$

and

$$\psi \circ \sigma = \text{identity on } W.$$

(This shows that any two sets sharing the universal property of $\displaystyle \bigtimes_{i \in I} A_i$ are related by a one-to-one and onto function, and justifies our attitude that any set having the universal property can be used as the Cartesian product of the family.)
(HINT: Construct σ and ψ similarly to h, and compute $\sigma \circ \psi$ and $\psi \circ \sigma$.)

(Solution is given on p. 40.)

2.4.3   Summary

The idea and notations (although possibly not the definitions) of families are straightforward and very useful. Many results would be difficult to write otherwise. The Cartesian product of a family is especially useful for constructing large mathematical structures from small ones.

The Cartesian product of a family is best thought of as

$$\bigtimes_{i \in I} A_i = \{f : I \longrightarrow \bigcup_{i \in I} A_i, \text{ where } \forall i \in I, f(i) \in A_i\}.$$

The ordered n-tuple approach can then be recovered by thinking of the index set as stretched out in a line, and associating $f(i)$ with the ith slot. The importance of the Cartesian product lies in the universal property that, for any set W which is equipped with functions $g_i : W \longrightarrow A_i$, there is a unique function

$$h : W \longrightarrow \bigtimes_{i \in I} A_i$$

which makes the following diagram commutative for each $i \in I$:

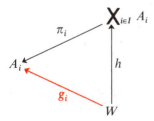

2.5 INVERSES AND COMPOSITES

2.5.0 Introduction

Inverse mappings and the composition of functions played such a large part in the Foundation Course that you must by now feel familiar with them. Halmos defines an inverse function in a quite different way. The other authors you will meet in this course define inverse function as in the Foundation Course, but one of them writes the composite function as $g \circ f$ where we would write $f \circ g$. The important thing, of course, is that all the authors are agreed on the essential nature of the basic ideas.

2.5.1 Notation and Terminology

*READ **Halmos**: page 38, line 1 to page 39, line 2.*

The embedded requests for proofs and examples appear as SAQ 16.

Notes

(i) *Halmos*: *page 38, lines 1 to 3.*
Compare this function with the function f_* described in Note (v) in Section 2.3.1. Note that $f_*(\varnothing) = \varnothing$.

(ii) *Halmos*: *page 38, lines 12 to 17.*
Halmos' inverse, f^{-1}, is not the *reverse mapping* of the Foundation Course, for that reverse mapping is not a function except when f is one-to-one, whereas, for Halmos, the inverse of *any* function is itself a function.

On the other hand, Halmos' use of *inverse image* does correspond to the use of *reverse image* in the Foundation Course.

(iii) *Halmos*: *page 38, line −5 to page 39, line 2.*
This "second interpretation", applicable only when f is one-to-one, is the only use of *inverse function* admitted by the Foundation Course and most other authors.

Halmos' approach rests on the principle that a function from X to Y is a special case of a relation from X to Y. To invert a relation, we merely reverse the order of the two elements in each ordered pair in the corresponding subset of the Cartesian product; hence we must do the same to invert a function. But what we obtain is not necessarily a function from Y to X. To make it a function, Halmos specifies a new domain, $\mathscr{P}(Y)$, and a new range, $\mathscr{P}(X)$. The rest of our authors prefer to retain X and Y and therefore forfeit the function property, except in the case when the correspondence is one-to-one.

SAQ 16

(i) Give the proof asked for, *Halmos*: *page 38, line 6.*

(ii) Give an example, as asked, *Halmos*: *page 38, line 7.*

(iii) Give the proof asked for, *Halmos*: *page 38, line −8.*

You may assume that f is a function with domain X. You are required to prove that the stated condition is satisfied if and only if f is *onto*.

(Solution is given on p. 41.)

2.5.2   Images and Inverse Images

*READ **Halmos**: page 39, lines 3 to −2.*

Notes

(i)   ***Halmos**: page 39, lines 5 to 19.*
Halmos here proves four simple results depending upon his definition of $f^{-1}(B)$. Since most authors use *inverse image* and $f^{-1}(B)$ in the same way, the four results hold for them too. (But note that, for most authors, contrary to Halmos, f^{-1} is not defined *as a function* unless f is one-to-one.)

Example

The second and fourth of the four results (*lines 9, 10 and 16, 17*) look as if they should give commutative diagrams. In fact, the two diagrams below are *not* correct. Can you redraw them correctly?

If f is onto, we have   If f is one-to-one, we have

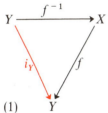   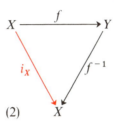

(1) Y (2) X

where i_Y is the identify map on Y.   where i_X is the identity map on X.

Solution

In both cases the error is the same: both diagrams involve Halmos' f^{-1}, and f^{-1} maps $\mathcal{P}(Y)$ to $\mathcal{P}(X)$, not Y to X. Since mappings cannot be composed unless the range of the first is included in the domain of the second, and $\mathcal{P}(A)$ is not included in A, we must draw the diagrams as follows:

f is onto: f is one-to-one:

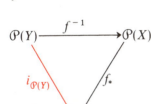   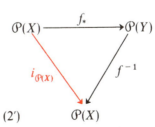

(1′) $\mathcal{P}(Y)$ (2′) $\mathcal{P}(X)$

replacing f by f_* (see Note (v), Section 2.3.1) and X and Y by $\mathcal{P}(X)$ and $\mathcal{P}(Y)$ respectively, with the consequent changes in the identity maps.

The following diagram shows an example of the left-hand diagram above and gives all the paths of all the elements, with

$X = \{a, b, c\}$, $Y = \{p, q\}$,

$f(a) = p$, $f(b) = f(c) = q$.

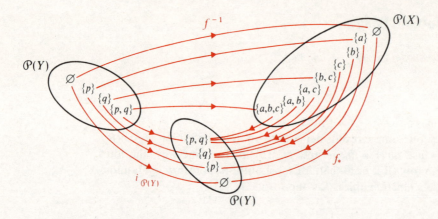

Notice that, because f is onto (although f^{-1} is not), the diagram is commutative for every element of $\mathcal{P}(Y)$, i.e.,

$$f_* \circ f^{-1} = i_{\mathcal{P}(Y)}.$$

We recommend that you draw the left-hand diagram for a case where f is not onto, say

$$X = \{a, b\}, \ Y = \{p, q\}, \ f(a) = f(b) = p,$$

and find an element in $\mathcal{P}(Y)$ for which the diagram is not commutative. Similarly, show all the paths for the right-hand diagram, first with

$$X = \{a, b\}, \ Y = \{p, q\}, \ f(a) = p, \ f(b) = q,$$

and verify that the diagram is commutative for all elements of $\mathcal{P}(X)$, and then take

$$X = \{a, b, c\}, \ Y = \{p, q\}, \ f(a) = p, \ f(b) = f(c) = q,$$

and find an element in $\mathcal{P}(X)$ for which the diagram is not commutative.

Now although diagrams (1) and (2) are undoubtedly wrong, and diagrams (1') and (2') are undoubtedly correct, it remains the fact that if either one of the following two diagrams is commutative, then the other will necessarily be commutative also, since the images of the elements under f, say, determine and are determined by the images of the elements under f_*.

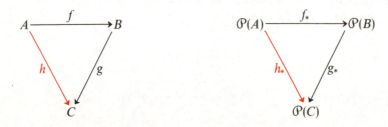

Further, changing sets to power sets may make the symbolism appear intolerably complicated. (See, for example, Note (ii) below.) Therefore, *in cases where nothing except the commutativity of a diagram is at issue*, we shall frequently abuse the notation and write X where, to be strictly correct, we should write $\mathcal{P}(X)$, and f where we should write f_*.

(ii) **Halmos**: *page 39, lines −16 to −12.*
This "unexceptionable" quality is a further justification of Halmos' definition of f^{-1}: it produces large numbers of "commutative" situations, which can equally well be represented as instances of distributivity or morphism. For example, *line −12* gives the following commutative diagram. (Note that f^{-1} is involved, so that all the sets should be power sets, but then the notation would be complicated and unwieldy. We have abused the notation as described in the previous note.)

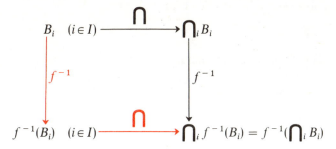

We have already seen (SAQ 16, (ii), page 26) that the corresponding result for f is false. It is precisely because f^{-1} possesses nice properties which f does not possess that topological definitions are given in terms of f^{-1}.

2.5.3 Composition of Functions

*READ **Halmos**: page 39, line -1 to page 40, line -13.*

The embedded exercise appears as part of SAQ 17.

Notes

(i) *Halmos: page 40, lines 7 to 11.*
Halmos and many other authors use gf where the Foundation Course used $g \circ f$. Another of the authors we shall meet reinstates the symbol $\circ$, although he writes $f \circ g$ where the others would write gf.

(ii) *Halmos: page 40, lines -16 to -13.*
It is very important to observe that

   $f^{-1}g^{-1}$ is the inverse of gf

is true with Halmos' interpretation of an inverse function. For the Foundation Course and many other authors it is true only if the functions are one-to-one. For most authors, two functions are inverses if and only if each of their composites is the identity function. It is false, with Halmos' interpretation of f^{-1}, that $f^{-1}g^{-1}gf$ is the identity. Indeed, with Halmos' interpretation, the composite $f^{-1}f$ is not necessarily the identity, as shown by the following example.

Let $X = \{a, b, c\}$, $Y = \{p, q\}$, and f be such that

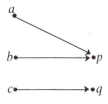

For instance, in this case,

   $f^{-1}f(a) = \{a, b\}$

and, since $\{a\} \neq \{a, b\}$, $f^{-1}f$ is not the identity function.

Note also that, strictly, for Halmos, f is not the inverse of f^{-1}. We need to involve f_* again. In other words, for Halmos, "is the inverse of" is not a symmetric relation on the set of functions.

SAQ 17

(i) Taking X as $\{a, b, c\}$, construct two one-to-one functions f and g from X onto itself such that $fg \neq gf$.

(ii) If σ maps S to T, τ maps T to U and μ maps U to V, prove that

$$\mu(\tau\sigma) = (\mu\tau)\sigma,$$

i.e., if the domains and ranges of functions are such that composition is possible, then composition is associative.

(Solution is given on p. 41.)

2.5.4 Inversion and Composition of Relations

READ *Halmos*: *page 40, line -12 to the end of the section on page 41.*

The embedded queries and request for a proof appear in SAQ 18 and SAQ 20.

Notes

(i) *Halmos*: *page 40, lines -11 to -9.*
Remember that a relation from X to Y is a subset of the Cartesian product $X \times Y$. If this subset is R, we can construct the set

$$R^{-1} = \{(a, b):(b, a) \in R\}.$$

Then R^{-1} is the inverse relation from Y to X. As "verbal" examples, on a set of men, the inverse of "is the father of" is "is a son of" while, on a set containing men and women, the inverse of "is the father of" is "is a son or daughter of the man".

(ii) *Halmos*: *page 40, line -2 to page 41, line 3.*
If R is the subset of the Cartesian product $X \times Y$ and S is the subset of the Cartesian product $Y \times Z$, we construct the set

$$SR = \{(x, z): \exists y \in Y \text{ such that } (x, y) \in R \text{ and } (y, z) \in S\},$$

and this is the composite relation from X to Z. Observe that, in general, $SR \neq RS$, even when both composite relations are defined, just as functional composition is not commutative.

(iii) *Halmos*: *page 41, lines 6 to 7.*
If you cannot see that SR means "is a nephew of" try saying that R "maps" men to their fathers and S "maps" men to their brothers.

(iv) *Halmos*: *page 41, lines 7 to 11.*
The R, S and T here are those at the beginning of the paragraph, except that R and S are now *functions* with X, Y and Z as before. It follows that T is a function from X to Z. Since a function is just a special type of relation, we should perhaps properly regard composition of functions as a special case of composition of relations.

(v) *Halmos*: *page 41, line 15.*
(Compare this remark with Note (ii) in Section 2.5.3.) We can construct

$$R^{-1}S^{-1}SR = (SR)^{-1}SR;$$

we shall, in general, find that it contains pairs other than pairs of the form (a, a), $a \in X$, so that it is not the identity relation on X (or on a subset of X).

(vi) *Halmos*: *page 41, lines 21 to 23.*
It can easily be seen that

$$I \subset R, \, R \subset R^{-1} \text{ and } RR \subset R$$

exactly correspond to the verbal definitions of the three properties given earlier. In some instances, the verbal definitions seem ambiguous: if so, the forms given here will be helpful. For example, using arrows in the standard way, the following diagram shows a relation in $X = \{a, b, c\}$. Is this relation transitive?

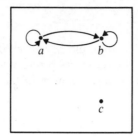

Solution

Here

$$X = \{a, b, c\}$$

and

$$R = \{(a, a), (b, b), (a, b), (b, a)\}.$$

We form

$$RR = \{(a, a), (a, b), (b, b), (b, a)\}.$$

Now $RR \subset R$, so R is transitive in X.

SAQ 18

(i)   Answer the four questions in **Halmos**: *page 41*, *line 7*, in the same context, and in the same manner, that Halmos dealt with the meaning of *SR*.

(ii) If R is a relation from X to Y and S is a relation from Y to Z, prove that

$$(SR)^{-1} = R^{-1}S^{-1}.$$

(iii) Answer the question in **Halmos**: *page 41, lines 20 and 21*.

(Solution is given on p. 42.)

2.6 SUMMARY OF THE TEXT

This text has been concerned mainly with concepts which should have been familiar (in an intuitive fashion) to students of the Foundation Course. Its primary purpose has been to show that it is possible to define all these concepts in terms of sets, based on set theory axioms.

The ideas introduced in the Foundation Course which are consolidated and perhaps extended in this text are the following:

> ordered pairs,
> relations, especially equivalence relations,
> equivalence classes* and partitions,
> functions or mappings*,
> the properties of operations on sets,
> inverse images*,
> inverse functions*,
> composition of functions*,
> connection between composition and inversion for functions and relations*.

New Ideas and Terms

The following new concepts have been introduced in this text:

> the range* of a function or a relation,
> the projection, inclusion, restriction and extension maps,
> the set of all functions from one set to another,
> the canonical map*,
> the characteristic function of a subset,
> families,
> the Cartesian product of a family*.

* The asterisks against certain items in the two lists above indicate that there are differences in the notation or terminology used in these topics by the various authors whom we have met or shall meet in this course. It is not necessary that you should know by heart the terminology or notation which an author uses, but it is *very* important that you should know the regions in which differences exist. It is then easy enough (and should become automatic), when referring to a book, to ascertain that particular author's usage.

Postscript

"... the present gigantic development of the mathematical faculty is wholly unexplained by the theory of natural selection, and must be due to some altogether distinct cause."

A. R. Wallace
Darwinism, Chapter 15

2.7 FURTHER SELF-ASSESSMENT QUESTIONS

SAQ 19

Decide which of the properties (reflexivity, symmetry, transitivity) are possessed by the following relations:

(i)   (ii)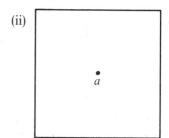

(iii)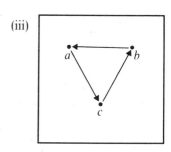

(Solution is given on p. 42.)

SAQ 20

Work ***Halmos***: *page 41*, the exercise at the bottom of the page, parts (i) and (ii).

(Solution is given on p.42.)

SAQ 21

Let $f : A \longrightarrow B, \qquad g : B \longrightarrow C$.

(i)   Show that, if gf is one-to-one, then f is one-to-one. Find a counter-example to the conjecture that g must be one-to-one.

(ii)   Show that, if gf is onto, then g is onto. Find a counter-example to the conjecture that f must be onto.

(Solution is given on p.43.)

SAQ 22

Show that A^A is closed under the binary operation $\circ$ (composition).

Show also that there is a member $\lambda \in A^A$ such that

$$f \circ \lambda = f = \lambda \circ f \qquad \forall f \in A^A.$$

(Solution is given on p. 44.)

2.8 SOLUTIONS TO SELF-ASSESSMENT QUESTIONS

Solution to SAQ 1

(i) $(a, b) \cup (c, d) = \{\{a\}, \{a, b\}\} \cup \{\{c\}, \{c, d\}\}$
$$= \{\{a\}, \{c\}, \{a, b\}, \{c, d\}\}$$

(ii) $(a, b) \cup (a, b) = \{\{a\}, \{a, b\}\} = (a, b)$

(iii) $(a, b) \cap (a, b) = \{\{a\}, \{a, b\}\} = (a, b)$

(iv) $(a, b) - (b, a) = \{\{a\}, \{a, b\}\} - \{\{b\}, \{a, b\}\} = \{\{a\}\} = (a, a)$

(v) $(a, b) - (a, a) = \{\{a\}, \{a, b\}\} - \{\{a\}\} = \{\{a, b\}\}$

Solution to SAQ 2

(i)   If $A = \{a\}$, $B = \{b\}$, then
$$A \times B = \{(a, b)\} = \big\{\{\{a\}, \{a, b\}\}\big\}$$
and
$$B \times A = \{(b, a)\} = \big\{\{\{b\}, \{a, b\}\}\big\}.$$
Thus
$$A \times B \neq B \times A.$$

(ii) No. The ordered pairs of $A \times (B \times C)$ are ordered pairs in which the second element is an ordered pair: those of $(A \times B) \times C$ have first elements which are ordered pairs. (Express each as a set if you like—it involves an awful lot of braces!)

(iii) Consider $R = \{(a, b), (c, d)\}$. If this were equal to $A \times B$, then we would have $A \supset \{a, c\}$ and $B \supset \{b, d\}$; hence $(a, d) \in A \times B$. But $(a, d) \notin R$, so $R \neq A \times B$.

Solution to SAQ 3

In each of (i), (ii) and (iii) we prove the equality of two sets by proving that each set is contained in the other.

(i)   Suppose $\alpha \in (A \cup B) \times X$.
Then $\alpha = (c, x)$, where $c \in A \cup B$, $x \in X$.
If $c \in A$, then $\alpha \in A \times X$, while if $c \in B$ then $\alpha \in B \times X$.
In either case, $\alpha \in (A \times X) \cup (B \times X)$, so
$$(A \cup B) \times X \subset (A \times X) \cup (B \times X). \qquad *$$

On the other hand, suppose $\beta \in (A \times X) \cup (B \times X)$.
Then either $\beta = (a, x)$, where $a \in A$, $x \in X$, or $\beta = (b, x)$, where $b \in B$, $x \in X$.
In either case, $\beta \in (A \cup B) \times X$, so
$$(A \times X) \cup (B \times X) \subset (A \cup B) \times X. \qquad **$$

From the two inclusions * and **, it follows that the two sets are equal.

(ii) Suppose $\alpha \in (A \cap B) \times (X \cap Y)$.
Then $\alpha = (c, z)$, where $c \in A \cap B$, $z \in X \cap Y$; that is, $c \in A$, $c \in B$, $z \in X$, $z \in Y$.
But $c \in A$ and $z \in X$ implies $\alpha \in A \times X$
and $c \in B$ and $z \in Y$ implies $\alpha \in B \times Y$,
so $\alpha \in (A \times X) \cap (B \times Y)$, whence
$$(A \cap B) \times (X \cap Y) \subset (A \times X) \cap (B \times Y). \qquad *$$

On the other hand, suppose $\beta \in (A \times X) \cap (B \times Y)$.
Then $\beta = (d, w)$, where $d \in A$, $w \in X$ and $d \in B$, $w \in Y$.
But $d \in A$ and $d \in B$ implies $d \in (A \cap B)$
and $w \in X$ and $w \in Y$ implies $w \in (X \cap Y)$:

hence $\beta \in (A \cap B) \times (X \cap Y)$, so that

$$(A \times X) \cap (B \times Y) \subset (A \cap B) \times (X \cap Y). \qquad \text{**}$$

From the two inclusions * and **, it follows that the two sets are equal.

(iii) Suppose $\alpha \in (A - B) \times X$.
Then $\alpha = (e, x)$, where $e \in A - B$, $x \in X$; that is, $e \in A$, $e \notin B$, $x \in X$.
Now $e \in A$, $x \in X$ implies $\alpha \in (A \times X)$
and $e \notin B$, $x \in X$ implies $\alpha \notin (B \times X)$,
so $\alpha \in (A \times X) - (B \times X)$ and therefore

$$(A - B) \times X \subset (A \times X) - (B \times X). \qquad \text{*}$$

On the other hand, suppose $\beta \in (A \times X) - (B \times X)$;
then $\beta = (f, x)$, where $f \in A$, $f \notin B$, $x \in X$.
Now $f \in A$, $f \notin B$ implies $f \in A - B$;
therefore, since $x \in X$, $\beta \in (A - B) \times X$, and therefore

$$(A \times X) - (B \times X) \subset (A - B) \times X. \qquad \text{**}$$

From the two inclusions * and **, it follows that the two sets are equal.

(iv) Every element of $A \times B$ is of the form $\{\{a\}, \{a, b\}\}$, where $a \in A$, $b \in B$. Now if either A or B is empty, there are no elements of this form, so $A \times B$ is empty.

Conversely, if neither A nor B is empty, then their Cartesian product is not empty; hence, if the Cartesian product is empty, then either A or B is empty.

(v) Suppose that $A \subset X$ and $B \subset Y$.
$X \times Y = \{x : x = (r, s), r \in X \text{ and } s \in Y\}$.
If $(p, q) \in A \times B$, then $p \in A$ and $q \in B$.
But since $A \subset X$ and $B \subset Y$, it follows that $p \in X$ and $q \in Y$; hence $A \times B \subset X \times Y$.
Conversely, suppose that $A \times B \subset X \times Y$, where $A \times B \neq \varnothing$, and $a \in A$.
Then, since $A \times B \neq \varnothing$, $B \neq \varnothing$, so we may choose *any* $b \in B$. Thus $(a, b) \in A \times B$, whence $(a, b) \in X \times Y$,
i.e., $a \in X$, $b \in Y$, and so $A \subset X$, $B \subset Y$.

Note: such results as these often give commutative diagrams; for example, (i) gives

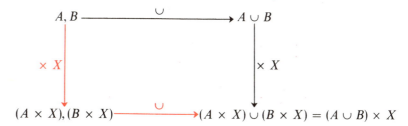

There is a similar diagram for (iii).

Solution to SAQ 4

The relation is inclusion between the elements of $\mathcal{P}(X)$.

Solution to SAQ 5

(i) R is symmetric and transitive but not reflexive. (Remember that the reflexive property requires $x R x$ for *every* $x \in X$.)

(ii) R is reflexive, symmetric and transitive.

Solution to SAQ 6

(The solutions given here are not unique.)

The following relation is symmetric and transitive but not reflexive on X:

$$R = \{(b, b), (c, c), (b, c), (c, b)\}.$$

This relation is illustrated in the following figure.

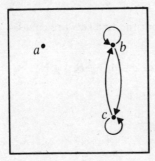

The following relation is reflexive and transitive but not symmetric:

$$R = \{(a, a), (b, b), (c, c), (a, b)\}.$$

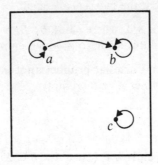

The following relation is reflexive and symmetric but not transitive:

$$R = \{(a, a), (b, b), (c, c), (a, b), (b, a), (b, c), (c, b)\}.$$

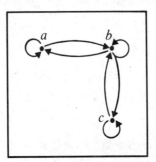

Solution to SAQ 7

(The following solution is not unique.)

Let $a \in X$. We define a/R to be

$$\{x \in X : xRa\} = \{x \in X : (x, a) \in R\}$$

and X/R to be

$$\{A \in \mathcal{P}(X) : \text{for some } a \in X, A = a/R\}.$$

Solution to SAQ 8

(a) The answer to this part depends upon our views about evolution: if we believe that every person in the world is descended from Adam and Eve, then the suggested relation is an equivalence relation of a trivial kind. (For there is only one equivalence class.)

On the other hand, consider the partial family tree shown in the following diagram, in which p and q are the parents of a, etc., and where we assume that no lines above those shown cross the dotted line. Then $a \sim b$ and $b \sim c$, but we do not have $a \sim c$, so the relation is not transitive, and therefore not an equivalence relation.

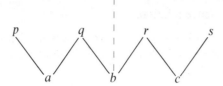

(b) This relation is not transitive.

If the residences A, B and C of a, b and c respectively are in a straight line; and $AB = 90$ miles, $BC = 90$ miles, then $a \sim b$ and $b \sim c$ but a is not related to c.

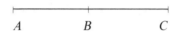

Therefore the relation is not an equivalence relation.

(c) This relation is an equivalence relation.

(d) This relation is an equivalence relation for which each class, except the class to which 0 belongs, contains just two elements.

(e) This relation *is* symmetric and transitive but it is not reflexive: hence it is not an equivalence relation.

(f) The answer here depends upon our convention on whether a line L is parallel to itself. If so, the relation is an equivalence relation. If not, reflexivity and transitivity fail.

Solution to SAQ 9

(a) The suggested proof begins "Let $a \sim b$" and ends, after perfectly sound argument, "then $a \sim a$". This does not, however, prove in general that the relation is reflexive, for reflexivity requires that *every* element of the set should be related to itself. This proof merely shows that, given the hypothesis, *if* an element is related to *some* element, *then* it is related to itself.

For example, given the set $\{a, b, c\}$, the assurance that the symmetric and transitive properties hold, and that $a \sim b$, the proof given will show that $b \sim a$, $a \sim a$ and $b \sim b$. It will *not* show that $c \sim c$ and hence it will not show the reflexive property. These assurances turn the partial diagram on the left of the following figure in to the complete diagram on the right: they do not produce a loop at c.

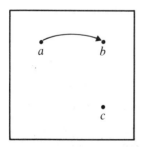 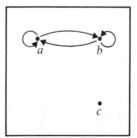

(b) For instance, if instead of

"for all a in A, $a \sim a$"

we take Property 1 to be

"for all $a \in A$ and some $b \in A$, $a \sim b$ implies $a \sim a$", then Properties 2 and 3 together will imply the new Property 1.

A relation possessing these properties will not necessarily be an equivalence relation. The set $A = \{a, b, c\}$ with

$$R = \{(a, b), (b, a), (a, a), (b, b)\}$$

is a counter-example. R is not an equivalence relation.

Notice that the reflexive property of an equivalence relation on a set X corresponds to that part of the definition of a partition which says that the union of the subsets is X (**Halmos**: *page 28, line 6*).

Solution to SAQ 10

There are $3^2(=9)$ functions in Y^X and $2^3(=8)$ functions in X^Y.

Solution to SAQ 11

The following diagram may help. It shows a dot for each element of the Cartesian product $\{a, b\} \times \{c, d, e\}$, where $X = \{a, b\}$, $Y = \{c, d, e\}$.

The domain of each projection here contains six elements: the projection of $X \times Y$ onto X maps three elements to each element of X, and the projection of $X \times Y$ onto Y maps two elements of $X \times Y$ to each element of Y. If either projection is to be one-to-one, it is plain, since the dot-pattern must always be rectangular, that at least one of the sets X and Y must be a singleton. In symbols, if p_X is the projection of $X \times Y$ to X,

$$p_X(x, y) = x \text{ for all } (x, y) \in \{(x, y) : y \in Y\},$$

so the condition for p_X to be one-to-one is that, for each $x \in X$, $\{(x, y) : y \in Y\}$ should be a singleton, i.e. that Y should be a singleton. Similarly, if X is a singleton, p_Y is one-to-one.

Solution to SAQ 12

(i) Every function from $\varnothing$ to Y involves a subset of $\varnothing \times Y$. Now whether Y is empty or not, $\varnothing \times Y$ is the empty set, because it has no members. We have to show that this is indeed a function.

Now if a set of ordered pairs does *not* correspond to a function, then there must exist either

(a) an element of the domain which does not appear as first element of some pair in the set of ordered pairs,

or

(b) an element of the domain which appears in the set of ordered pairs as first member of at least two distinct pairs.

But neither (a) nor (b) is possible in this case, since the domain contains no elements. Hence $\varnothing \times Y$ defines a unique function from $\varnothing$ to Y.

(ii) Every function from X to $\varnothing$ corresponds to a subset of $X \times \varnothing$. Now $X \times \varnothing$ is empty, so the only possible function from X to $\varnothing$ has the empty set as its set of ordered pairs. But X is not empty. Let $a \in X$. Now there is no pair with first element a in the (empty) set of ordered pairs; hence there are no functions from X to $\varnothing$.

Solution to SAQ 13

(a) χ_E maps every $x \in E$ to 1 and every other x in A to 0.
χ_F maps every $x \in F$ to 1 and every other x in A to 0.
Now the product $\chi_E(x)\chi_F(x)$ is 1 if and only if

$$\chi_E(x) = 1 \text{ and } \chi_F(x) = 1,$$

that is, $x \in E$ and $x \in F$, so $x \in E \cap F$.

If $x \notin E \cap F$, then either $\chi_E(x)$ or $\chi_F(x)$ (or both) is 0, so the product is 0.

Hence $\chi_E \cdot \chi_F = \chi_{E \cap F}$.

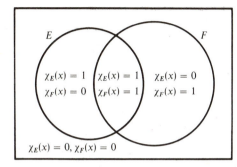
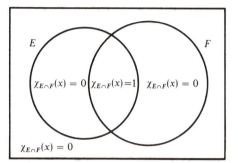

(b) By part (a),

$$\chi_E + \chi_F - \chi_{E \cap F} = \chi_E + \chi_F - \chi_E \cdot \chi_F.$$

We can use this to tabulate the values of $(\chi_E + \chi_F - \chi_E \cdot \chi_F)(x)$ for all possible values of $\chi_E(x)$ and $\chi_F(x)$:

$\chi_E(x)$	$\chi_F(x)$	$\chi_E \cdot \chi_F(x)$	$(\chi_E + \chi_F - \chi_E \cdot \chi_F)(x)$
0	1	0	1
0	0	0	0
1	1	1	1
1	0	0	1

We see that the entry in the last column is 1 when either $\chi_E(x)$ or $\chi_F(x)$ (or both) is 1, and the entry is 0 only if $\chi_E(x)$ and $\chi_F(x)$ are both 0. Thus x has image 1 if and only if x is in either E or F or both. Hence the given function is the characteristic function of $E \cup F$.

In the terminology of the Foundation Course, (b) shows that the function χ mapping sets to their characteristic functions is a morphism of sets under union to the combination of functions by $*$, where

$$a * b = a + b - ab.$$

We have the following commutative diagram:

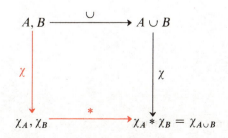

Under this morphism, $E \cup F \longmapsto \chi_E * \chi_F$, so

$$(E \cup F) \cup G \longmapsto (\chi_E * \chi_F) * \chi_G.$$

Now $\cup$ is associative, so $*$ is associative whence

$$E \cup F \cup G \longmapsto \chi_E * \chi_F * \chi_G \,;$$

hence,

$$\chi_{E \cup F \cup G} = \chi_E + \chi_F + \chi_G - \chi_E \cdot \chi_F - \chi_F \cdot \chi_G - \chi_E \cdot \chi_G + \chi_E \cdot \chi_F \cdot \chi_G,$$

giving, by repeated application of the result in part (a),

$$\chi_{E \cup F \cup G} = \chi_E + \chi_F + \chi_G - \chi_{E \cap F} - \chi_{F \cap G} - \chi_{E \cap G} + \chi_{E \cap F \cap G}.$$

An alternative solution for part (b) is as follows. Take $\chi'_E(x) = 0$ for $x \in E$ and $\chi'_E(x) = 1$ for $x \in E'$; that is,

$$\chi'_E(x) = 1 - \chi_E(x).$$

Then

$$\chi_{E \cup F}(x) = \chi'_{(E' \cap F')}(x) \qquad \text{(by De Morgan's law)}$$

$$= 1 - (1 - \chi_E(x))(1 - \chi_F(x))$$

and hence

$$\chi_{E \cup F \cup G}(x) = 1 - (1 - \chi_E(x))(1 - \chi_F(x))(1 - \chi_G(x)).$$

Solution to SAQ 14

Let $S = \bigcup_i A_i$: then $Z \subset (S^I)$. Notice that we need a definition of $I \times S$ in order to define S^I, so that this general Cartesian product is based on the Cartesian product of two sets.

Solution to SAQ 15

For each i we have

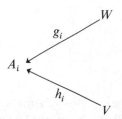

By the universal property of W, and that of V, we know that there are unique functions

$$\sigma : W \longrightarrow V \qquad \text{and} \qquad \psi : V \longrightarrow W$$

such that the respective diagrams are commutative. Further, for any $w \in W$, $v \in V$,

$$\sigma(w) \in V \qquad \text{and} \qquad \psi(v) \in W.$$

Now each member of V and W is associated with a function from I to $\bigcup_{i \in I} A_i$. Therefore we can specify $\sigma(w)$ and $\psi(v)$ by

$$\sigma(w) : i \longmapsto g_i(w) \qquad \text{and} \qquad \psi(v) : i \longmapsto h_i(v).$$

Now we examine $\sigma \circ \psi$ and $\psi \circ \sigma$.

$$\sigma \circ \psi(v) : i \longmapsto g_i(\psi(v))$$

$$= \psi(v)(i)$$

$$= h_i(v)$$

$$= v(i).$$

Thus $\sigma \circ \psi$ is the identity on each function $v \in V$. Similarly,

$$\psi \circ \sigma(w): i \longmapsto h_i(\sigma(w))$$
$$= \sigma(w)(i)$$
$$= g_i(w)$$
$$= w(i).$$

Thus $\psi \circ \sigma$ is the identity on each function $w \in W$.

There is therefore a one-to-one onto function between V and W.

Solution to SAQ 16

(i) We use f to represent the mapping of elements, and f_* to represent the mapping of subsets. Now

$$f_*(X) = \bigcup_{x \in X} f_*(\{x\}).$$

$$y \in f_*\left(\bigcup_{i \in I} A_i\right) \Leftrightarrow \exists x \in \bigcup_{i \in I} A_i \text{ such that } y = f(x)$$
$$\Leftrightarrow \exists j \in I, \exists x \in A_j \text{ such that } y = f(x)$$
$$\Leftrightarrow \exists j \in I \text{ such that } y \in f_*(A_j)$$
$$\Leftrightarrow y \in \bigcup_{i \in I} f_*(A_i).$$

Therefore

$$f_*\left(\bigcup_{i \in I} A_i\right) = \bigcup_{i \in I} f_*(A_i).$$

(ii) Let $X = \{a, b, c\}$, $Y = \{p, q\}$, $A_1 = \{a, b\}$, $A_2 = \{b, c\}$ and let $f(a) = f(c) = q$, $f(b) = p$.
Then $\bigcap_i A_i = \{b\}$ and $f_*(\bigcap_i A_i) = \{p\}$.
But $\bigcap_i f_*(A_i) = \{p, q\}$, and $\{p, q\} \neq \{p\}$. Thus

$$f_*\left(\bigcap_i A_i\right) \neq \bigcap_i f_*(A_i).$$

However, it is true that

$$f_*\left(\bigcap_i A_i\right) \subset \bigcup_i f_*(A_i).$$

(iii) For every $y \in Y$, if $f^{-1}(\{y\})$ is non-empty, there is at least one element $x \in X$ such that $f(x) = y$: that is, f is *onto* Y.

Conversely, if f maps X *onto* Y, then every element $y \in Y$ is the image $f(x)$ of some $x \in X$: hence the inverse image of every $y \in Y$ is non-empty.

Thus the condition is necessary and sufficient.

Solution to SAQ 17

(i) A suitable example (but not, of course, the only one) is the following:

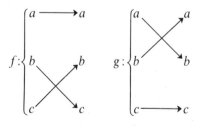

for then

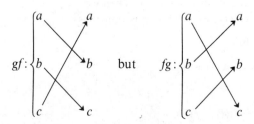

(ii) This is LEMMA 1.1 on **Herstein**: *page 14*: the proof is given there. Notice that Herstein writes $\sigma \circ \tau$ where Halmos would write $\tau\sigma$ and the Foundation Course would write $\tau \circ \sigma$. Also, where the Foundation Course and Halmos would write $\sigma(x)$, Herstein writes $x\sigma$. (Notice how little the change of convention matters.)

Solution to SAQ 18

(i) (a) R^{-1} means "is the father of".
  (b) S^{-1} means "is the brother of" ($=S$).
  (c) RS means "is the son of" ($=R$).
  (d) $R^{-1}S^{-1}$ means "is the uncle of" ($=(SR)^{-1}$).

(ii) We use the most common method; that is, we prove that each set is contained in the other.
Let $(z, x) \in (SR)^{-1}$, where $z \in Z$ and $x \in X$.
Then, by the definition of $(SR)^{-1}$, $(x, z) \in SR$ and there must be an element y in Y such that $(x, y) \in R$ and $(y, z) \in S$.
Hence $(y, x) \in R^{-1}$ and $(z, y) \in S^{-1}$ so that

$$(z, x) \in R^{-1}S^{-1}. \qquad\qquad *$$

Secondly, let $(z, x) \in R^{-1}S^{-1}$. It follows that there is a y in Y such that $(z, y) \in S^{-1}$ and $(y, x) \in R^{-1}$.
Hence $(y, z) \in S$ and $(x, y) \in R$, so $(x, z) \in SR$ and

$$(z, x) \in (SR)^{-1}. \qquad\qquad **$$

Thus, from $*$ and $**$, we see that each element of one set is an element of the other: hence the sets are equal.

(iii) The connections are:

  (a) if ran R = dom $R^{-1} = X$, then $I \subset RR^{-1}$;
  (b) if dom R = ran $R^{-1} = X$, then $I \subset R^{-1}R$.

Proofs are given below.

  (a) Let $r \in$ ran R. Then $(a, r) \in R$ for some a and $(r, a) \in R^{-1}$.
  Hence $(r, r) \in RR^{-1}$, by the definition of RR^{-1}. Now $I = \{(r, r) : r \in X\}$.
  Hence, if $r \in$ ran R for all $r \in X$, i.e. ran $R = X$, then $I \subset RR^{-1}$.
  (b) Let $s \in$ dom R. Then $(s, a) \in R$ for some a and $(a, s) \in R^{-1}$.
  Hence $(s, s) \in R^{-1}R$ by the definition of $R^{-1}R$.
  Now, again, $I = \{(s, s) : s \in X\}$. Hence if $s \in$ dom R for all $s \in X$, i.e. dom $R = X$, then $I \subset R^{-1}R$.

Solution to SAQ 19

(i) This relation is reflexive, symmetric, transitive.

(ii) This relation is symmetric and transitive, but not reflexive.

(iii) This relation is not reflexive, not symmetric, not transitive.

Solution to SAQ 20

(i) We have $gf(x) = x$ for all $x \in X$. To show that f is one-to-one, suppose $f(x) = f(y)$. Then certainly $gf(x) = gf(y)$, and since gf is the identity, we conclude that $x = y$. This means that f must be one-to-one. To show that g is onto, let $x \in X$. Then we know that $gf(x) = x$, so putting $y = f(x)$, we have $g(y) = x$. Thus g is onto.

(ii) There are two parts here. We need to show first that

if $\forall A, B \subset X, f(A \cap B) = f(A) \cap f(B)$, then f is one-to-one,

and, secondly, that

if f is one-to-one, then $\forall A, B \subset X, f(A \cap B) = f(A) \cap f(B)$.

To show the first part, the harder of the two, we choose A and B carefully. Suppose

$f(x) = f(y)$, but $x \neq y$.

Let $A = \{x\}$, $B = \{y\}$. Then

$f(A \cap B) = f(\varnothing) = \varnothing.$

But

$f(A) \cap f(B) = \{f(x)\} \neq \varnothing,$

a contradiction. Thus

$f(x) = f(y)$ means $x = y$, so f is one-to-one.

For the second part, we observe that certainly, $\forall A, B \subset X$,

$f(A \cap B) \subset f(A) \cap f(B).$ *

Suppose $x \in f(A) \cap f(B)$. Then $\exists a \in A$, $\exists b \in B$ such that

$f(a) = x = f(b).$

But f is one-to-one, so $a = b$, and hence $x \in f(A \cap B)$, so

$f(A) \cap f(B) \subset f(A \cap B).$ **

From the two inclusions, * and **, it follows that

$\forall A, B \subset X \quad f(A \cap B) = f(A) \cap f(B).$

Solution to SAQ 21

We have $f : A \longrightarrow B$, $g : B \longrightarrow C$.

(i) gf is one-to-one.

Let $x, y \in A$, and suppose that

$f(x) = f(y).$

Then

$gf(x) = gf(y)$

and by the one-to-one-ness of gf,

$x = y,$

so f is one-to-one.

To illustrate the fact that g need not be one-to-one, take $A = \{a\}$, $B = \{b_1, b_2\}$, $C = \{c\}$ and let

$f : a \longmapsto b_1$

$g : b_1 \longmapsto c$

$g : b_2 \longmapsto c.$

Then

$gf : a \longmapsto c$ is one-to-one, but g is not.

(ii) gf is onto.

To show that g is onto, let $c \in C$.
We need to find $b \in B$ such that $g(b) = c$.
Now gf is onto, so there is an $a \in A$ with

$gf(a) = c.$

Thus

$g(f(a)) = c,$

and $f(a) \in B$, so $f(a)$ is the required element of B.

The previous example illustrates that f need not be onto.

Solution to SAQ 22

Let $f, g \in A^A$.

Since the codomain of f is $A = $ domain g, $g \circ f$ is a function with domain A and co-domain A, and so $g \circ f \in A^A$. To find λ such that $f \circ \lambda = f = \lambda \circ f$, we observe that λ cannot alter any element of A, and so the most likely candidate is the identity function:

$$\lambda : a \longmapsto a \qquad (a \in A).$$

The diagrams

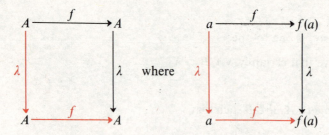

are certainly commutative, and the composition in both cases is just f, as required.

Unit 3 Sets and Numbers

"Symbolism is useful because it makes things difficult. Now in the beginning everything is self-evident, and it is hard to see whether one self-evident proposition follows from another or not. Obviousness is always the enemy to correctness. Hence we must invent a new and difficult symbolism in which nothing is obvious . . ."

Bertrand Russell,
International Monthly, 1901, p. 85.

Contents

Set Books

P. R. Halmos, *Naive Set Theory*, paperback edition 1972 (Van Nostrand Reinhold).
I. N. Herstein, *Topics in Algebra*, paperback edition (Xerox College/T.A.B.S, 1964).
B. Mendelson, *Introduction to Topology*, paperback edition 1972 (Allyn and Bacon).
M. L. Minsky, *Computation: Finite and Infinite Machines*, paperback edition 1972 (Prentice-Hall).

It is essential to have these books; the course is based on them and will not make sense without them.

Unit 3 is based on **Halmos**, Sections 11–17.

Conventions

Before working through this correspondence text make sure you have read *A Guide to the Course: Topics in Pure Mathematics*.

References to the Open University Mathematics Foundation Course Units (The Open University Press, 1971) take the form *Unit M100 3, Operations and Morphisms*.

3.0 INTRODUCTION

This unit is based on Sections 11 to 17 inclusive of **Halmos**. First, we put the definition of the natural numbers on an axiomatic footing, using the properties and axioms of set theory that we introduced in *Units 1* and *2*. We then define recursively the usual arithmetic operations of addition, multiplication and exponentiation, and introduce the idea of *order* into the set of natural numbers. Finally, we survey briefly the Axiom of Choice and some related ideas.

You will probably find this unit difficult, representing, as it does, an unfamiliar approach to very familiar material. However, the approach to the natural numbers adopted in the text is typical of the axiomatic method so often used in pure mathematics, and is worth mastering.

The final section, on the Axiom of Choice, involves what are perhaps the most difficult concepts of the entire course. At this point we present only a very brief introduction to these ideas.

Aims

After working through this text, you should be able to:

(i)   understand the construction of the set of natural numbers as a minimal successor set, the construction being justified by the Axiom of Infinity;

(ii)   recognize the Peano axioms, and, in particular, the Principle of Mathematical Induction, as derivable from the construction in (i);

(iii) appreciate the part played by the Recursion Theorem in definition by induction and, in particular, in defining the operations of addition, multiplication and exponentiation on the natural numbers;

(iv)   understand the role of one-to-one correspondence in defining equivalent sets and in distinguishing finite and infinite sets;

(v)   appreciate the meanings of the technical terms used in defining partially ordered and totally ordered sets and in describing possible additional properties of such sets (in particular, "least," "greatest," "minimal," "maximal," "infimum," "supremum");

(vi)   enjoy at least a superficial acquaintance with the Axiom of Choice, Zorn's Lemma and the Well Ordering Principle, and see that the last has significant implications in connection with the Principle of Transfinite Induction.

3.1 NUMBERS

3.1.0 Introduction

In the previous two units we introduced some basic ideas concerning sets. However, if we wish to do arithmetic, we must have numbers. As the counting process is very fundamental in life, one might take the viewpoint which says that we all know what the natural numbers are, so let us not bother with any nonsense about defining them. But do we really know what "two" is? In any case, it is just the fact that the arithmetical process is so fundamental which makes it worth our while to take some time over the definition of the natural numbers. We have seen in the Foundation Course (*Unit M100 34, Number Systems*) that we can construct the integers, rationals and real numbers starting from the natural numbers; so if we define the natural numbers in terms of the set concepts developed in the earlier sections of **Halmos** we shall have reduced "all arithmetic to sets". This part of the commentary will be based on **Halmos**, *Section 11*.

3.1.1 Numbers

*READ **Halmos**: page 42, line 1 to page 44, line 13.*

Note

The important fact to emphasize is that we are defining the natural numbers in terms of sets. In this context, the symbol 0 is just an alternative way of writing the unique set $\emptyset$ which has no members. The symbol 1 is, by definition, the set 0^+.

$$1 = 0^+ = 0 \cup \{0\} = \emptyset \cup \{\emptyset\} = \{\emptyset\}.$$

The symbols $0, 1, 2, 3, \ldots$ are well known and traditional. (We assume that Halmos is using base ten (or more) for this representation because of his use of 956 in the last sentence.) Of course, there are numerous equivalent alternative specifications of the (later) natural numbers in terms of sets. For example,

$$3 = 2^+ \quad \text{or} \quad 2 \cup \{2\} \quad \text{or} \quad \{0, 1, 2\}$$

or, in terms of $\emptyset$ only,

$$3 = \big\{\emptyset, \{\emptyset\}, \{\emptyset, \{\emptyset\}\}\big\}.$$

The "usual" notation, 3, is the most concise.

*READ **Halmos**: page 44, line 14 to page 45, line 4.*

Notes

(i)   **Halmos**: *page 44, lines 17 and 18*.

The Axiom of Infinity. This extremely powerful axiom is the "deeper and more useful existential assumption" referred to previously, **Halmos**: *page 8, line 5*. (It actually contains at least two axioms and two results, but our set theory looks neater if we do not overburden it with axioms.) First of all, recall that so far there is no axiom guaranteeing the existence of a set; a temporary assumption to that effect was made in **Halmos**: *Section 3* pending the present Axiom of Infinity. We now have, as an official sub-axiom of the Axiom of Infinity:

Sub-Axiom 1: There exists a set.

The same reasoning as in **Halmos**: *Section 3* leads to

Result 1: There is the empty set $\emptyset$.

Although at this point we do not officially know what "finite" means, without being too pompous, it is clear that, using $\emptyset$, we can construct some finite sets.

For example, a set with one member, $\{\emptyset\}$, or one with three members, $\{\emptyset, \{\emptyset\}, \{\{\emptyset\}\}\}$. (These sets are not necessarily natural numbers.) Therefore we have:

Result 2: There are finite sets.

The final sub-axiom contained in this powerful principle is

Sub-Axiom 2: There is a successor set.

As we saw in Section 2 of **Halmos**, it is not sufficient to write down a sentence to specify a set, and we need at least one *successor set A*. On the basis of our previous axioms we could not have claimed that the "thing" containing 0 and the successor of each of its elements was, in fact, a set.

(ii)   **Halmos**: *page 44, lines* -12 *to* -10.

Are there successor sets other than ω? Such a set must have ω as a subset, and, if it contains any other element, a say, then it must also contain $a^+, a^{++}, \ldots$. (For example, a might be the set $\{\{\emptyset\}\}$.) However, the axioms that we have at present do not guarantee that this set exists! On the other hand, the axioms do not explicitly forbid the existence of such a set. In fact, in Section 19, Halmos does construct a successor set different from ω, but he needs another axiom to do so.

Example
(**Halmos**: *page 44, line* -11).

Show that the intersection of every non-empty family of successor sets is a successor set.

Preamble

We give some advice here about proof in general. Before tackling any proof, it is worth while setting out exactly what one has to prove. This involves Step 0:

(0)   ESTABLISH THE NOTATION
The notation may be suggested by the problem or, as is the case here, may need to be devised by the solver.

We then proceed to Step 1:

(1)   Detail the RESULT TO BE PROVED, listing any sub-results which need proof.

Finally, Step 2:

(2)   Proceed to PROVE all results in (1).

In our example, the steps are as follows:

(0)   ESTABLISH THE NOTATION
     As there is no notation suggested by the statement of the problem, we must devise our own. We shall take $\{S_{i \in I}\}$ as our family of successor sets, indexed by $i \in I$, where I is a non-empty index set.

(1)   RESULT TO BE PROVED
     We are asked to prove that the intersection of the family of successor sets is a successor set, that is, $S = \bigcap_i S_i$ is a successor set. (Note that at this stage it is not unusual to introduce further notation (such as S) in order to simplify matters.) This entails proving the following sub-results:

     (i) S is a set;

     (ii) S is a successor set.

     The second result (ii) means that we must demonstrate two facts:

     (iia) 0 belongs to S;

     (iib) if x belongs to S, then x^+ belongs to S.

Now that the steps have been set out in such detail, we find, as is often the case, that the result proves itself.

(2) PROOF
  (i) S is a set, since the intersection of any non-empty family of sets is a set.
        (See **Halmos**: *page 15, lines 18–34*.)

  (iia) Since each S_i is a successor set, by definition $0 \in S_i$, for every i.
        Therefore $0 \in \bigcap_i S_i$, and so $0 \in S$.

  (iib) Let $x \in S$.
        Then $x \in S_i$, for every $i \in I$.
        Since each S_i is a successor set,

$$(x \in S_i) \Rightarrow (x^+ \in S_i) \qquad \text{for each } i.$$

        Since $x^+ \in S_i$ for every i, $x^+ \in \bigcap_i S_i$.
        Therefore $x^+ \in S$.

As we have demonstrated all the required results (i), (iia) and (iib), the proof is complete.

Following Halmos, we define $\omega = \bigcap_i A_i$, where the intersection is taken over *all* the successor sets contained in A. By the previous example, ω is a successor set. However, at this point it may not be clear that, if we started with a different successor set C, and formed

$$\omega' = \bigcap_i C_i$$

for all the successor sets C_i contained in C, then

$$\omega = \omega'.$$

Example

Prove that $\omega = \omega'$.

Preamble

As for the previous example, we again set out the steps necessary to solve this problem.

(0) ESTABLISH THE NOTATION
    This has been done in the statement of the problem and in the preceding paragraph.

(1) RESULT TO BE PROVED
    We must show that

    (i) $\omega = \omega'$.

    You will recall that almost all such proofs of equalities between two sets are split into two parts:

    (ia) $\omega \subset \omega'$;

    (ib) $\omega' \subset \omega$.

    We shall adopt this technique in the proof.

(2) PROOF
    (ia) We show that $\omega \subset \omega'$.
          Since ω' is a successor set, as the previous example demonstrated, we can replace the set B of **Halmos**: *page 44, line -9 to line -6* by ω', so that $\omega \subset \omega'$.

    (ib) We wish to show that $\omega' \subset \omega$.
          Clearly, since A and C are successor sets, so too is $A \cap C$, and since $A \cap C \subset A$, it follows that $A \cap C$ is a successor set contained in A, so that, by the definition of ω,

$$\omega \subset A \cap C.$$

Further, since $\omega \subset A \cap C \subset C$, the set ω is a successor set contained in C, and, by the definition of ω', $\omega' \subset \omega$.

Since we have established (ia) and (ib), the proof is complete.

At this point it should be clear that the argument can be considerably abbreviated. Halmos has shown that, if *B is an arbitrary successor set*, then a minimal successor set (like ω or ω') is a subset of *B*. Since both ω and ω' are successor sets, each can play the role of *B*, so that

$\omega \subset \omega'$ (where $\omega' = B$, and ω is minimal)

$\omega' \subset \omega$ (where $\omega = B$, and ω' is minimal).

Therefore

$\omega = \omega'$ and ω is unique.

(iii) **Halmos**: *page 44, line* -3.
Some authors (e.g. Halmos) include 0 in "the natural numbers": 0 was excluded in *Unit M100 34, Number Systems*.

READ **Halmos**: *page 45, line 5 to the end of Section 11.*

Note

(i) **Halmos**: *page 45, line 15.*
When Halmos refers to a family $\{x_i\}$ whose index set is a natural number, he is using the notation of *page 34, the first paragraph of Section 9*. Consider, for example, a family $\{x_i\}$ whose index set is the natural number 3. That is, we have a function $x: I \longrightarrow X$, where the index set I is 3. Therefore the family $\{x_i\}$, where $i \in I$, consists of exactly three members, namely x_0, x_1 and x_2, since $I = 3 = \{0, 1, 2\}$; or, more accurately, the image set of the function x contains exactly three images, x_0, x_1 and x_2.

SAQ 1

In the text it is stated that all possible unordered pairs, $\{a, b\}$, with $a \neq b$, do not constitute a set (**Halmos**: *page 42, line* -6). Why not?

(Solution is given on p. 30.)

SAQ 2

Show that $3 \quad 2 \neq 1$.

(Remember that the only interpretation of   allowed here is that used in set complementation.)

(Solution is given on p. 30.)

SAQ 3

Given that *A* and *B* are successor sets, show that

(i) $A \cup B$ is a successor set;

(ii) $A - B$ is *not* a successor set;

(iii) $\mathcal{P}(\omega)$ is *not* a successor set.

(Solution is given on p. 30.)

3.2 THE PEANO AXIOMS

3.2.0 Introduction

We have given an axiomatic description of the set of natural numbers, based on the set theory so far presented. The object of this section is to show the equivalence of the system so constructed and that defined by the axioms of Peano. Halmos writes down each of the five axioms of Peano, and shows that they are consequences of our definition of the set ω. This part of the commentary is based on *Halmos*, *Section 12*.

3.2.1 Deduction of the Peano Axioms

*READ **Halmos**: page 46, line 1 to page 47, line 13.*

Notes

(i) *Halmos*: page 46.
The statements (I), (II), (III), (IV) and (V) are the Peano Axioms. We wish to prove that the set ω fulfils each of them. Then, since we have already constructed ω, we need not regard the statements (I) to (V) as *axioms*, but *consequences* of our construction. The definition of ω as the minimal successor set has (I), (II) and (III) as immediate consequences.

Axiom (III) of Peano, the Principle of Mathematical Induction, is worth spending a little time on, as it provides us with an indispensable mathematical tool which we shall use repeatedly in theorems involving the natural numbers. You have, of course, met the method called "proof by induction" in *Unit M100 17*, *Logic II*: wishing to prove that some statement, $P(n)$ say, involving an arbitrary natural number n, is true for all $n \in \omega$, we show that:

(1) $P(0)$ is true;

(2) if $P(n)$ is true, then $P(n^+)$ is true.

We can now establish the validity of this proof by defining the set

$S = \{n \in \omega : P(n) \text{ is true}\}$.

(1) tells us that $0 \in S$ and (2) tells us that, if $n \in S$, then $n^+ \in S$. Axiom (III) enables us to conclude that $S = \omega$ and so $P(n)$ must be true for all $n \in \omega$.

We can provide a further illustration of the use of this axiom in the following example, where we deduce from it a second principle of induction, sometimes called "course of values induction".

Example
If $T \subset \omega$, if $0 \in T$, and if $n^+ \in T$ whenever $0, 1, 2, \ldots, n$ are all members of T, then $T = \omega$.

PROOF
We are going to use the Principle of Induction and, as with almost all proofs of this sort, it is absurdly simple once we define the appropriate set S. Let

$S = \{n \in \omega : 0, 1, 2, \ldots, n \text{ are all members of } T\}$.

Clearly $0 \in S$, and, if $n \in S$, then $0, 1, 2, \ldots, n$ are all members of T. But this implies that $n^+ \in T$. It follows that $n^+ \in S$, since $0, 1, 2, \ldots, n, n^+$ all belong to T.

We have shown that

$n \in S$ implies $n^+ \in S$

so, by the Principle of Induction, $S = \omega$. It follows from the definition of S that $S \subset T$, so

$\omega = S \subset T \subset \omega$.

Hence

$$T = \omega.$$

Many proofs by induction are greatly simplified by the use of course of values induction, because, instead of having to prove that $n^+ \in S$ knowing only that $n \in S$, we can now assume that all of $0, 1, 2, \ldots, n$ are in S. In Section 3.3.2, Note (vii), we shall use this form of induction in an example.

You will also meet a generalization of this principle in **Halmos**: *page 66*, where the concept of transfinite induction is introduced.

(ii) **Halmos**: *page 47, line 9 to line 11.*
In the following example, we show that the two alternative definitions of a transitive set given in this passage are equivalent.

Example

Prove that the following properties of a set E are equivalent:

(i)   if $x \in y$ and $y \in E$, then $x \in E$;

(ii) E includes ($\subset$) everything that it contains ($\in$), i.e. if $z \in E$, then $z \subset E$.

Solution

We first show that (i) $\Rightarrow$ (ii).

Let $z \in E$. Then, if $t \in z$, we have

$$t \in z \text{ and } z \in E \text{ and hence } t \in E,$$

by property (i). That is, all the elements of z are elements of E and so $z \subset E$.

We next show that (ii) $\Rightarrow$ (i).

Let $x \in y$ and $y \in E$.

Then $y \subset E$, by property (ii), which means that $x \in E$.

This completes the proof.

An example of a transitive set which is not a natural number is

$$E = \{\varnothing, \{\varnothing\}, \{\{\varnothing\}\}\}.$$

Examples of sets that are *not* transitive are

$$E' = \{\varnothing, \{\{\varnothing\}\}\},$$

because $\{\varnothing\} \in \{\{\varnothing\}\}$ and $\{\{\varnothing\}\} \in E'$ but $\{\varnothing\} \notin E'$,

and

$$E'' = \{1, 2, 3\},$$

because $\varnothing \in 1$ and $1 \in E''$ but $\varnothing \notin E''$.

READ **Halmos**: *page 47, line 14 to page 48, line 5.*

Note

The first three paragraphs constitute the proof of Axiom (V). Since the proof is somewhat tortuous, and the result is exactly what one would expect, you should simply read over the section, comforted by the thought that the proof is there. *You will not be asked to reproduce it.*

3.2.2   The Recursion Theorem

This section is very important. In it we show how to give a set-theoretic interpretation of the process of definition by a recursion formula, or definition by induction. This process is used very often in mathematics and we shall draw attention to it repeatedly in this course.

READ Halmos: page 48, line 6 to the end of the section (omitting the Exercise for the moment).

General Comment

The following diagram illustrates definition by induction.

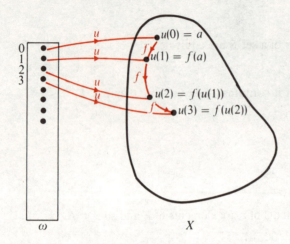

The corresponding commutative diagrams are as follows:

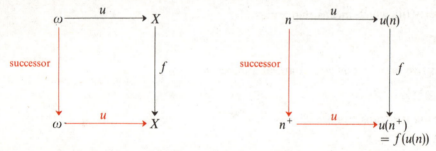

(The "successor" function maps each element of ω to its successor.)

The diagram on the right simply makes the process explicit by choosing a specific $n \in \omega$.

We can make the diagram more revealing as follows.

Introduce an arbitrary set containing a single element (we shall use $\{\varnothing\}$ for this set) and represent the element a in X as a function a from $\{\varnothing\}$ to X, where $a(\varnothing) = a$. Similarly, represent the element 0 in ω as a function 0 from $\{\varnothing\}$ to ω where $0(\varnothing) = 0$.

We then state the result as follows.

For *any* X, for *any* $f : X \longrightarrow X$ and for *any* a in X, there is *just one* $u : \omega \longrightarrow X$ such that the following diagram is commutative.

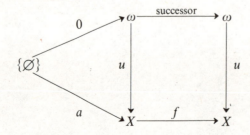

12

At a first reading, the proof of the Recursion Theorem may seem a little complicated, so we shall spend some time on it.

We wish to construct a function u from ω to X which does the job, that is,

$u(0) = a,$

and

$u(n^+) = f(u(n))$ for all $n \in \omega$.

Now all functions from ω to X are subsets of $\omega \times X$. Halmos considers the collection $\mathcal{C}$ of those subsets A of $\omega \times X$ for which $(0, a) \in A$, and

$(n, x) \in A \Rightarrow (n^+, f(x)) \in A$.

That is,

$\mathcal{C} = \{A \in \mathcal{P}(\omega \times X): (0, a) \in A \text{ and } (n, x) \in A \Rightarrow (n^+, f(x)) \in A\}$.

The intersection $u = \bigcap_{A \in \mathcal{C}} A$ belongs to $\mathcal{C}$. Halmos now proves by induction that u is a *function* with domain ω. (See SAQ 4.) Since u is, by definition, the "smallest" member of $\mathcal{C}$, it is unique.

The above process of definition (of u) may have a familiar ring about it. It is a common method of defining uniquely the "smallest" set of elements A for which a given condition $S(A)$ holds:

Let $\mathcal{C}$ = collection of sets A for each of which $S(A)$ holds;
show that $\mathcal{C}$ is non-empty;
show that $\bigcap \mathcal{C}$ is such that $S(\bigcap \mathcal{C})$ holds, i.e. $\bigcap \mathcal{C} \in \mathcal{C}$;
then $\bigcap \mathcal{C}$ is the "smallest" set of elements A for which $S(A)$ holds.

Of course, for the method to work, $\mathcal{C}$ must be a set, and $\bigcap \mathcal{C}$ must indeed have the required property. For example, this was the method used in **Halmos**: *Section 11* to define ω, where A was a successor set, i.e., $S(A)$ was the condition

$0 \in A$ and $x \in A \Rightarrow x^+ \in A$.

In the present case, the sets of $\mathcal{C}$ are analogous to the successor sets and the resulting function u is analogous to ω.

The following SAQ is intended to be helpful. We have written out the statement and proof of the Recursion Theorem, but have left some gaps (boxes) for you to fill in. We suggest that you test your understanding of this theorem by attempting the question *without referring to* **Halmos**. If, however, you have had difficulty in understanding the proof in **Halmos**, then cheat! Enter our solutions (given on p. 31) in the boxes and then try to understand this version of the proof. It is exactly the same as Halmos' proof, but we have set out the steps in more detail. (Because each step is carefully enunciated, the proof appears rather long and complicated—it is not as bad as it looks.)

If you do not find the SAQ helpful, then ignore it! However, make sure that you understand the statement of the theorem; you will come across applications of this result several times in this course, particularly in the units based on **Minsky**.

SAQ 4

Fill in the boxes in the following statement and proof of the Recursion Theorem; the letters in the boxes refer to the solution.

THEOREM

The Recursion Theorem states that, if a is an element of a set X, and if f is a function from X into X, then there exists a function u from ω into X such that $u(0) = a$, and

$u(n^+) = \boxed{}^{A}$ for all n in ω.

PROOF

Any function u from ω to X is a subset of $\boxed{}^{B}$.

Consider the collection $\mathcal{C}$ of all subsets A of $\omega \times X$ for which $(0, a) \in A$, and for which

$(n, x) \in A \Rightarrow \boxed{}^{C} \in A$.

Since $\boxed{}^{D} \in \mathcal{C}$, $\mathcal{C}$ is not empty.

Form the intersection $u = \bigcap_{A \in \mathcal{C}} A$.

We shall show that $u \in \mathcal{C}$. In order to prove this, we have to prove two things:

(i) $(0, a) \in u$;

(ii) $(n, x) \in u \Rightarrow \boxed{}^{E} \in u$.

PROOF OF (i)

Since $(0, a) \in A$ for each $A \in \mathcal{C}$, we have $(0, a) \in \bigcap_{A \in \mathcal{C}} A$, that is,

$(0, a) \in \boxed{}^{F}$.

PROOF OF (ii)

$(n, x) \in u \Rightarrow (n, x) \in A$ for each $A \in \mathcal{C}$

$\Rightarrow (n^+, f(x)) \in A$, for each $A \in \mathcal{C}$

$\Rightarrow \boxed{}^{G} \in u$.

We must now show that u is a *function*; that is, if $(n, x) \in u$ and $(n, y) \in u$, then

$x = \boxed{}^{H}$.

We also wish to show that the domain of u is ω, so, altogether, we should like to show that, for each $n \in \omega$, $(n, x) \in u$ for exactly one x.

We prove this by induction. Let

$S = \{n \in \omega : (n, x) \in u \text{ for exactly one } x\}$.

It is clear that we must prove that $S = \omega$.

In order to prove this, we must again show two things:

(i) $0 \in S$;

(ii) $n \in S \Rightarrow n^+ \in S$.

PROOF OF (i)

Let us suppose that, on the contrary, $0 \notin S$.

Since $(0, a) \in u$, and we know that $0 \notin S$, then, by the definition of S, there must be

another element $b \in X$ such that $a \neq b$ and $(0, b) \in \boxed{}^{I}$.

Consider the set $u - \{(0, b)\}$. Now this diminished set still contains $(0, a)$, since $a \neq b$. Further, it has the property that, if it contains (n, x), it also contains $(n^+, f(x))$; for u had that property, and the discarded element $(0, b)$ is not equal to $(n^+, f(x))$ for any n,

since $0 \neq \boxed{}^{J}$.

Therefore $u - \{(0, b)\} \in \mathcal{C}$, which contradicts the construction of u as the smallest set of $\mathcal{C}$. Our hypothesis that $0 \notin S$ is false, and therefore $0 \in S$.

PROOF OF (ii)

Suppose now $n \in S$.

This means that there exists a unique element $x \in X$ such that $\boxed{}^{K} \in u$.

Since $(n, x) \in u$, it follows that $\boxed{}^{L} \in u$.

Suppose that $n^+ \notin S$; then we must have $(n^+, y) \in u$, where $y \neq \boxed{}^{M}$.

Consider the diminished set $\bar{u} = u - \{(n^+, y)\}$.

We shall exhibit a contradiction by showing that $\bar{u}$, which is *properly* contained in u, is a member of $\mathcal{C}$.

In order to show that $\bar{u} \in \mathcal{C}$, we must yet again prove two things:

(a) $(0, a) \in \bar{u}$;

(b) $(m, t) \in \bar{u} \Rightarrow \boxed{}^{N} \in \bar{u}.$

PROOF OF (a)

$(0, a) \in \bar{u}$, since $(0, a) \in u$ and $0 \neq \boxed{}^{O}$.

PROOF OF (b)

Let $(m, t) \in \bar{u}$.

First case: If $m = n$, then $(n, t) \in \bar{u} \Rightarrow (n, t) \in u$, since $\bar{u} \subset \boxed{}^{P}$.

Since $n \in S, t = \boxed{}^{Q}$.

Therefore $(n^+, f(x)) \in u$; and since $f(x) \neq y, (n^+, f(x)) \in \bar{u}$, that is, $(m^+, f(t)) \in \bar{u}$.

Second case: If $m \neq n$, then $m^+ \neq n^+$, and so $(m^+, f(t)) \in \bar{u}$.

This completes the proof of (b).

This shows that $\bar{u}$, which is smaller than u, belongs to $\mathcal{C}$, which is a contradiction.

And so our assumption that $n^+ \notin S$ was false. Therefore $n^+ \in \boxed{}^{R}$.

This completes the proof of (ii).

By the Principle of Mathematical Induction, $S = \omega$.

This completes our proof that u is a function from ω to X.

(Solution is given on p. 31.)

SAQ 5

Prove that, if n is a natural number, then $n \neq n^+$.

(Solution is given on p. 31.)

SAQ 6

Prove that, if $n \in \omega$ and $n \neq 0$, then $n = m^+$ for some natural number m.

HINT: Use induction on the set

$$S = \{n \in \omega : n = m^+ \text{ for some } m \in \omega, \text{ or } n = 0\}.$$

(Solution is given on p. 31.)

SAQ 7

Prove that, if E is a non-empty subset of some natural number, then there exists an element k in E such that $k \in m$ whenever m is an element of E distinct from k.

(Solution is given on p. 31.)

SAQ 8

Describe the action of the function u of the Recursion Theorem in the following cases:
(i)   the function f is the identity map $i : X \longrightarrow X$, and $u(0) = a$;
(ii)   the set X is $B \times B$;

the function $f : B \times B \longrightarrow B \times B$ is defined by

$$f(b_1, b_2) = (b_2, b_1),$$

and

$$u(0) = (a_1, a_2).$$

(Solution is given on p. 31.)

3.3 ARITHMETIC

3.3.0 Introduction

Now that we have defined the natural numbers, it is natural to want to add, subtract, and multiply them. We have not as yet given any definitions of these operations. It is the task of this section to define these operations in set-theoretic terms, and show that the results agree with our intuitive ideas.

Since the natural numbers have been defined as sets, one is tempted to look first at the usual set operations in order to define addition, say. The obvious candidate, union, does not work. For example, $2 \cup 1 = 2$, not 3. However, we already have $2^+ = 3$, and so perhaps we could use the successor function $+ : n \longmapsto n^+$ as the function $f : X \longrightarrow X$ of the Recursion Theorem (in our case, the function $+ : \omega \longrightarrow \omega$). This is the clue which enables us to use recursion to define addition, and, subsequently, multiplication and exponentiation. We shall develop these ideas in Section 3.3, in which the ideas of order and the number of elements of a finite set are also introduced.

3.3.1 Sums, Products and Powers

READ Halmos : page 50, line 1 to page 51, line 10.

Notes

(i) **Halmos**: *page 50, line 5.*
Halmos defines a "+" here (in "$m + n$") other than the raised $^+$ which he uses for "successor" and which is distinct from the "+" defined on *page 18, line 16* (symmetric difference).

(ii) **Halmos**: *page 50, lines 2–5.*
In order to see clearly how the Recursion Theorem has been applied, we need a dictionary to translate the terms used previously into those used now. We replace X by ω, f by the successor function, and a by m; the function u which results is now named s_m. Our dictionary reads as follows:

Recursion Theorem	*Definition of Sum Function*
$a \in X$	$m \in \omega$
$f : X \longrightarrow X$	$+ : \omega \longrightarrow \omega$, i.e., $+ : n \longmapsto n^+ \qquad (n \in \omega)$
$u : \omega \longrightarrow X$	$s_m : \omega \longrightarrow \omega$
$u(0) = a$	$s_m(0) = m$
$u(n^+) = f(u(n))$	$s_m(n^+) = (s_m(n))^+$

The diagram on page 12 of this commentary becomes the following, in which we show the action on 0 and on a general element n of ω.

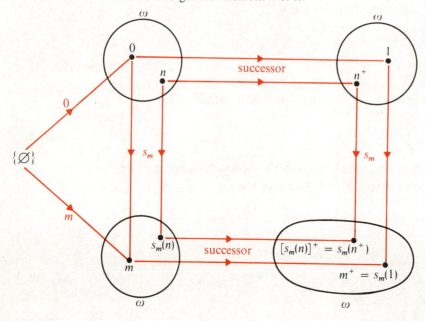

16

(iii) **Halmos**: *page 51, line 8.*

Note the abuse of notation in using m^n for $e_m(n)$; this notation has a different but not unrelated meaning on **Halmos**: *page 30, line* -10.

We shall not go into the rather tedious, but straightforward, formal proofs that addition and multiplication as here defined are commutative and associative. The following examples should suffice to convince you that the functions s_m, p_m and e_m map $n \in \omega$ to $m + n$, $m \times n$ and m^n respectively, and so their properties should come as no surprise.

SAQ 9

Let (m, n) have the values $(0, 0)$, $(0, 1)$, $(1, 0)$, $(1, 1)$, $(1, 2)$, $(2, 1)$, $(2, 2)$. Use the fact that

$$s_m(0) = m \quad \text{and} \quad s_m(n^+) = (s_m(n))^+$$

to calculate $s_m(n)$ for the above values of m and n, and verify that in those cases the "intuitive" result $m + n$ is obtained. Hence verify the commutativity and associativity of addition in these cases.

(Solution is given on p. 32.)

SAQ 10

In the statement of the Recursion Theorem, take $X = \omega$, a as 0 and $f: X \longrightarrow X$ as s_m, i.e. $f: \omega \longrightarrow \omega$, where $f(n) = m + n$. Then the function u is p_m.

Compute $p_m(n)$ for the same (m, n) values as in SAQ 9, verifying that in those cases $p_m(n)$ has the familiar value $m \times n$, and hence verify commutativity and associativity of multiplication.

(Solution is given on p. 32.)

SAQ 11

In the statement of the Recursion Theorem, take $X = \omega$, a as 1 and $f: X \longrightarrow X$ as p_m, i.e. $f: \omega \longrightarrow \omega$, where $f(n) = m \times n$. Then the function u is e_m.

Evaluate $e_m(n)$ for the same values of m and n as in SAQ 9, verifying that the "intuitive" result m^n is obtained. Show that exponentiation is neither commutative nor associative as a closed binary operation in ω.

(Solution is given on p. 32.)

SAQ 12

Make a dictionary, as in Note (ii) of this section, to translate the terms of the Recursion Theorem into the language defining the product function.

(Solution is given on p. 32.)

SAQ 13

Make a dictionary to translate the terms of the Recursion Theorem into the notation used in defining the exponent function.

(Solution is given on p.32.)

SAQ 14

We have used the Recursion Theorem repeatedly to define new operations in ω. Having defined addition in ω, we used it to define *multiplication* in ω by taking

$$f: n \longmapsto m + n = s_m(n), \text{ with } a = 0$$

and then used multiplication to define *exponentiation* in ω by taking

$$f: n \longmapsto m \times n = p_m(n), \text{ with } a = 1.$$

Continue the process for one more step, taking $a = 1$ and choosing f in the obvious way, to define a new operation $*$ on ω by $m * n = u(n)$. Is $*$ commutative? Associative?

(Solution is given on p. 33.)

3.3.2 Order in the Natural Numbers

Although defined as sets, the natural numbers have been seen to acquire most of the attributes expected of them. There remains only the interpretation of $m < n$ or $m \leqslant n$, that is, the question of *ordering* the natural numbers.

READ Halmos: page 51, line 11 to page 52, line 13.

Notes

(i)　*Halmos: page 51, line 19.*
The Principle of Mathematical Induction is used repeatedly in the proof of the Assertion; firstly, to show that $S(0) = \omega$, and, secondly, to show that, if $S(n) = \omega$, then $S(n^+) = \omega$. However, the result is of more importance than the proof, which you will not be asked to reproduce.

(ii)　*Halmos: page 51, lines −6 and −5.*
The proof of the Assertion being complete, we now know that any two natural numbers m and n are comparable; that is, *at least* one of the following holds:

　　　(i) $m \in n$　　(ii) $m = n$　　(iii) $n \in m$.

In fact, *exactly* one holds. For example, suppose that

　　　(i) $m \in n$　　and　　(ii) $m = n$.

From (i), m is an element of n, and, from (ii), $n \subset m$, so that n is a subset of m. (From (ii) we also have $m \subset n$, but that is not relevant here.) Now (i) and (ii) together tell us that the natural number n is a subset of one of its elements m. This contradicts *Halmos: page 47, line 7, Statement (i)*, so this means that (i) and (ii) cannot hold simultaneously. Similarly, (ii) and (iii) cannot hold simultaneously for the same reason (the roles of m and n are reversed in the argument). Finally, (i) $m \in n$ and (iii) $n \in m$ cannot hold simultaneously. For (i) tells us that $m \subset n$, since every element of a natural number is a subset of it (*Halmos: page 47, line 8*), and then (iii) tells us that n is an element of m, so that m is a subset of one of its elements n, which is impossible (*Halmos: page 47, line 7*).

(iii)　*Halmos: page 52, lines 5 and 6.*
This is the more familiar notation. For any two natural numbers m and n, exactly one of the following holds:

　　　(i) $m < n$　　(ii) $m = n$　　(iii) $n < m$.

Alternatively, we say that one or both of the following hold:

　　　(i) $m \leqslant n$　　(ii) $n \leqslant m$.

If both the latter hold, then $m = n$.
(We shall see in the next section that the preceding statements mean that ω is a *totally ordered set* with respect to the order relation $\leqslant$.)

It is not always the case that any two elements of a set are "comparable" or, more generally, that a set with an order relation is totally ordered, as the following example illustrates.

Consider the set $\omega \times \omega$.

Define the relation ⊗ by $\forall (m_1, m_2), (n_1, n_2) \in \omega \times \omega$,

　　$(m_1, m_2) ⊗ (n_1, n_2)$ if $m_1 \leqslant n_1$ and $m_2 \leqslant n_2$.

In this case we should say that two elements a and b of $\omega \times \omega$ were "comparable" if

　　　either $a ⊗ b$　　or $b ⊗ a$　　(or both).

However, it is clear that not every two elements of $\omega \times \omega$ are comparable in this sense. For example, $a = (1, 2)$ and $b = (2, 1)$ are not comparable.

(iv)　*Halmos: page 52, line 7.*
In the other notation of relations,

　　$m < n$ means that $(m, n) \in <$,

where $< \subset \omega \times \omega$.

(See *Halmos: page 27, lines 1 to 5.*)

(v) **Halmos**: *page 52, Exercise (first part)*.

We prove here that

if $m < n$, then $m + k < n + k$, for all $k \in \omega$.

This seems clearly a place to use induction. Therefore, given m and n such that $m < n$, we define

$$S = \{k \in \omega : m + k < n + k\}.$$

We must prove that $S = \omega$. In order to prove this, we have to prove two things:

(i) $0 \in S$

(ii) $k \in S \Rightarrow k^+ \in S$.

It seems clear that (i) is easy to prove; but for (ii) we are going to need something like the following result:

if $m < n$ then $m^+ < n^+$.

This tells us how we should structure our proof: we prove the above result first.

LEMMA

If $m < n$, then $m^+ < n^+$.

PROOF OF LEMMA

We are told that $m \in n$. Now exactly one of the following holds:

(a) $m^+ \in n^+$ (b) $m^+ = n^+$ (c) $n^+ \in m^+$.

(b) is clearly false. For, by Axiom (V) of **Halmos**: *page 46*, this would imply that $m = n$, which contradicts the hypothesis.

(c) is false. For $n^+ \in m^+ \Rightarrow n^+ \in m$ or $n^+ = m$. Now $n^+ \in m$ and $m \in n$ leads to $n^+ \in n$ and so $n^+ \subset n$, by the transitivity of n (**Halmos**: *page 47, line 8*), and this is false, since the natural number n^+ cannot be a subset of one of its elements n (**Halmos**: *page 47, line 7*). Similarly, if $n^+ = m$, then $n^+ \in n$. Then, n^+, being an element of n, is a subset of n (**Halmos**: *page 47, line 8*), i.e., $n^+ \subset n$, which is false, as we have already remarked. So (c) is false.

Therefore only (a) can hold: $m^+ \in n^+$.

The proof of the main result is now straightforward.

PROOF

(i) Clearly $m + 0 < n + 0$ (since $m < n$ and $m + 0 = m$, $n + 0 = n$), so $0 \in S$.

(ii) Let $k \in S$, so that $m + k < n + k$.

By the Lemma,

$$(m + k)^+ < (n + k)^+.$$

Therefore

$$m + k^+ < n + k^+,$$

where we have used result (ii) of **Halmos**: *page 50, line −9*, and the commutativity of addition. Therefore $k^+ \in S$.

From (i) and (ii), it follows that $S = \omega$.

This completes the proof.

(vi) **Halmos**: *page 52, Exercise (second part)*.

The proof is similar to the proof given above: we do not give it here.

(vii) **Halmos**: *page 52, Exercise (third part)*.

This important theorem, proved in the following example, is a statement of the *Well Ordering Principle* for the natural numbers.

Example

Prove that, if E is a non-empty set of natural numbers, then there exists an element k in E such that $k \leqslant m$ for all m in E.

Preamble

The result seems "obvious" from at least two aspects:

(a) Does $0 \in E$? If so, then 0 is the required k in E.

    If $0 \notin E$, does $1 \in E$? If so, then 1 is the required k in E.

    ... and so on ...

(b) It is "obvious" that the intersection of a non-empty set of natural numbers is a natural number, and this intersection has the minimal property that we have often used before.

Unfortunately, it is difficult to make legitimate the "and so on" in (a) and equally difficult to prove the "obvious" in (b). We fall back on proof by contradiction. That is, we assume that there is *some* non-empty set E of natural numbers which does *not* have a least element, and hope to prove that this assumption leads to a contradiction and is therefore false, so that every non-empty set of natural numbers *does* have a least element.

(Note our conventional use of *least* element: if a set X has an element a such that $a \leqslant x$ for all $x \in X$, we say that a is the least element of X.)

If E has no least element, then certainly $0 \notin E$. But if $0 \notin E$, then certainly $1 \notin E$ or it would be the least element. Similarly, knowing that 0 and 1 are not in E, we can deduce that $2 \notin E$. This suggests that we can use induction to prove that E is empty and thus arrive at a contradiction. In a proof by induction, we show that some set S is equal to ω. What shall we choose for the set S? We want to show that E is empty, so we let

$$S = \omega - E.$$

Certainly $0 \in S$.

If $n \in S$, can we deduce that $n^+ \in S$? We cannot, but, if we know that $0, 1, 2, \ldots, n$ were all in S, then it would follow that $n^+ \in S$, for otherwise n^+ would be the least element of E. But this last statement is just what is needed for "course of values" induction (see our Example on page 10), and we can therefore conclude that $S = \omega$. Hence $E = \varnothing$, as required.

SAQ 15

Let E and F be two non-empty sets of natural numbers such that $E \subset F$.

Show that $\min F \leqslant \min E$, where $\min F$, $\min E$ are the least elements of F and E respectively.

(Solution is given on p. 33.)

3.3.3 Equivalent Sets and Finiteness

Historically, the natural numbers are associated with the important process of "counting". We shall see in this section how this process can be formalized, and we shall use the natural numbers, defined set-theoretically, to enumerate the elements of finite sets. Some of the results involving the number of elements of sets may seem rather obvious; it is unnecessary at this point to construct formal proofs of all the results indicated in this section; you should concentrate on understanding the concepts.

READ Halmos: page 52, line 14 to the end of the section, omitting the exercises and proofs of results for the present.

Notes

(i) *Halmos: page 52, line 15.*
Two sets E and F are *equivalent* if there is a function $f : E \longrightarrow F$ which is one-to-one and *onto*. This property is actually symmetric between E and F; for we may in accord with *Halmos: page 38, lines -7 to -1*, define a one-to-one onto function $f^{-1} : F \longrightarrow E$ by

$$\forall b \in F \quad f^{-1}(b) = a \text{ when } f(a) = b.$$

(Note that f^{-1} is used here in Halmos' second interpretation.)

We may therefore express the fact that E is equivalent to F by stating that the following two diagrams are commutative.

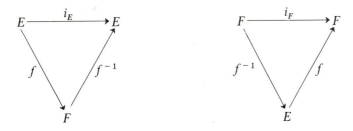

(Here i_E and i_F are the respective identity maps in E and F.)

More concisely, but less symmetrically, we have:

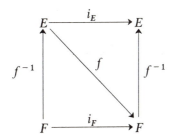

(ii) *Halmos: page 52, lines 20 to 27.*
It is easiest to see what is going on here by crudely writing out n, and taking an example for E, a proper subset of $n^+ = n \cup \{n\}$. Thus,

$$n = \{0, 1, 2, \ldots, (n-1)\}$$
$$n^+ = \{0, 1, 2, \ldots, (n-1), n\}$$
$$E = \{1, 2, n\} \qquad \text{(for example).}$$

In our case $n \in E$. Let us choose $k = 5$ as a number in n but not in E. Define the function $f : E \longrightarrow n$ by

$$f(1) = 1, f(2) = 2, f(n) = 5.$$

Then the image of E under f is a proper subset of n, which, by the induction hypothesis, is equivalent to some smaller natural number. Explicitly,

$$f(E) = \{1, 2, 5\}$$

and

$$f : E \longrightarrow \{1, 2, 5\} \text{ is one-to-one and } onto,$$

so that

$$E \sim \{1, 2, 5\}.$$

By the induction hypothesis,

$$\{1, 2, 5\} \sim \text{ some element of } n,$$

so that, by the *transitivity* of $\sim$,

$$E \sim \text{ some element of } n \text{ (i.e. of } n^+\text{)}.$$

(iv) **Halmos**: *page 53, line 5. Solution of Exercise.*

RESULT TO BE PROVED

The set ω is not equivalent to any natural number.

We shall prove this by contradiction. That is, we shall assume that ω is equivalent to some natural number n, and deduce a contradiction from this.

PROOF

Suppose ω is finite, that is, there is an $n \in \omega$ such that there exists a function $f : \omega \longrightarrow n$ which is one-to-one and onto.

Clearly for any one-to-one, onto function,

$$f : X \longrightarrow Y,$$

the restriction of f to A, a subset of X,

$$f|A : A \longrightarrow f(A)$$

is one-to-one and maps A onto a subset of Y. In our case, since $n^+ \subset \omega$, the function

$$f|n^+ ; n^+ \longrightarrow f(n^+)$$

is a one-to-one function which maps n^+ onto a subset of n. Therefore n^+ is equivalent to a proper subset of n^+, which is impossible according to **Halmos**: *page 52, line −6.*

Therefore ω is *not* finite.

(v) **Halmos**: *page 53, line 12. Solution of Exercise.*

We have only to display a one-to-one correspondence between ω and a proper subset of itself. There are many such, and we choose the one already given on **Halmos**: *page 52, line −10.*

The function $f : \omega \longrightarrow \omega - \{0\}$ defined by

$$f : n \longmapsto n^+$$

is clearly one-to-one and onto, and so

$$\omega \sim (\omega - \{0\}).$$

Therefore ω is equivalent to a proper subset of itself and so is not finite.

(vi) **Halmos**: *page 53, final paragraph of text.*

The results stated here concerning the numbers of elements of finite sets are fairly obvious. For the present it is unnecessary to construct formal proofs of these results. When Halmos states that the function

$$\# : S \longrightarrow \omega$$

(where S is the subset of $\mathcal{P}(X)$ comprising the finite subsets of X), is pleasantly related to the familiar set-theoretic relations and operations, he means that, in certain circumstances, we have commutative diagrams. It follows that morphisms can be set up. For example, if E and F are disjoint finite sets, we have:

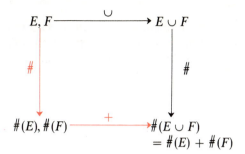

To make the following diagram a morphism, we need to define a domain closed under Cartesian product.

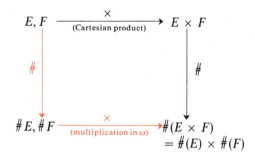

Similarly, the result

$$E \subset F \Rightarrow \#(E) \leqslant \#(F)$$

is an example of

$$a\rho_1 b \Rightarrow f(a)\rho_2 f(b),$$

where $\#$ plays the role of f, $\subset$ that of ρ_1 and $\leqslant$ that of ρ_2. This is an example of a morphism on a structure with a relation. (See *Unit M100 36, Mathematical Structures.*)

SAQ 16

Verify the statement of **Halmos**: *page 52, lines 16 and 17.*

(Solution is given on p. 33.)

SAQ 17

If E and F are finite sets, show that $E - F$ is finite, and that

$$\#(E \cup F) = \#(E - F) + \#(F).$$

(Solution is given on p. 34.)

SAQ 18

If E and F are finite sets such that $E \subset F$, show that

$$\#(E) \leqslant \#(F).$$

(Solution is given on p. 35.)

3.4 ORDER

3.4.0 Introduction

In Section 3.3.2 we discussed order in the set of natural numbers. We now examine the generalization of this idea. An introduction can be found in *Unit M100 19, Relations*, Section 19.3, which covers most of the points made by Halmos and gives examples of the many definitions.

3.4.1 Ordered Sets

READ Halmos: page 54, line 1 to page 57, line 7.

Notes

(i) *Halmos: page 54, lines 12 to 20.*
Most people think of $\leqslant$ in connection with the real numbers, where it is a *total* order: although it is also, of course, a partial order on the real numbers, its use as the standard symbol for any partial order on any set may prove confusing.

(ii) *Halmos: page 54, Solution of Exercise.*

R is anti-symmetric if and only if $R^{-1} \cap R \subset I$, where I is the identity relation,

$$I = \{(x, y) \in X \times X : x = y\}.$$

If R is a total order in X, then it is, of course, also a partial order and must on that account satisfy:

$$R \supset I \qquad \text{(reflexivity)}$$
$$R \circ R \subset R \qquad \text{(transitivity)}$$
$$R^{-1} \cap R \subset I \qquad \text{(anti-symmetry)}$$

and total order requires, further,

$$R^{-1} \cup R = X \times X.$$

(iii) *Halmos: page 56, line -11.*
A least element of an ordered set X is *unique*; for if a and b are least elements of X, $a \leqslant x$ and $b \leqslant x$, for all $x \in X$. Thus $a \leqslant b$ and $b \leqslant a$, whence $a = b$ (anti-symmetry). Note that, if a is a least element of X, then $a \in X$, by definition.

3.4.2 Infimum and Supremum

Although the concepts of *infimum* and *supremum* are placed in a very general context in this section, you may be more familiar with these concepts as related to R, the set of real numbers. The set R is, of course, a totally ordered set with respect to the usual order, $\leqslant$. It may help to fix your ideas if you construct examples based on R to illustrate the text. The only numbers so far available to Halmos are, of course, the natural numbers.

READ **Halmos**: page 57, line 8 to the end of Section 14.

Notes

(i) **Halmos**: *page 57, lines 10 to 12.*
Consider the set $E \subset \omega$ defined by

$$E = \{n \in \omega : n > 2\}.$$

This set has no upper bounds in ω and four lower bounds, 0, 1, 2 and 3, one of which belongs to E. Note that a bound of a set E in X need not belong to E.

(ii) **Halmos**: *page 57, lines 12 to 16.*
For the example of Note (i),

$$E_* = \{0, 1, 2, 3\}, E^* = \varnothing, E_* \cap E = \{3\}, \text{ a singleton.}$$

(iii) **Halmos**: *page 57, line 24.*
Again referring to the example of Note (i), E_* has as greatest element 3 and so 3 is inf E. On the other hand, E^* has no least element (having no elements at all) so E has no sup.

(If you are constructing examples for yourself using the real numbers as your set, note that the set is rather special: *every non-empty subset of R which has an upper bound has a supremum.* The set of rational numbers Q, for example, does not have this property. See *Unit M100 34, Number Systems.*)

SAQ 19

Consider the relation S in $\omega \times \omega$ as defined by **Halmos**: *page 57, line -1 (ii) to page 58, line 5,* and the subset $E = A \times A$, where $A = \{0, 1, 2\}$.

(i) Is the order total in $\omega \times \omega$?

(ii) Draw an "arrow diagram" representing the elements of E by points, with an arrow from (a, b) to (x, y) if and only if $(a, b)S(x, y)$.

(Since the relation is reflexive on $\omega \times \omega$, every point should really have a "loop" arrow to itself: the diagram will be less confusing if all such loops are taken for granted and not drawn in.)

(iii) Has E an inf in $\omega \times \omega$?

(iv) Has E a sup in $\omega \times \omega$?

(v) Is E_* finite in $\omega \times \omega$?

(vi) Is E^* finite in $\omega \times \omega$?

(vii) Consider also the subset F of $\omega \times \omega$,

$$F = \{(x, x) : x \in \omega\}.$$

Has F a least element? Has F a greatest element?

(Solution is given on p. 35.)

SAQ 20

Consider the collection $\mathcal{C}$ of all non-empty subsets of a non-empty set X, ordered by inclusion. Show that $(\mathcal{C}, \subset)$ is a partially ordered set, that each singleton is a minimal element of $\mathcal{C}$, but that $\mathcal{C}$ has no least element unless X itself is a singleton.

(**Halmos**: *page 57, line 1.*)

(Solution is given on p. 35.)

SAQ 21

Let $\mathcal{C}$ be the collection of all non-empty subsets of ω. Is $\mathcal{C}$ partially ordered with respect to the relation $\preccurlyeq$, where $A, B \in \mathcal{C}$, $A \preccurlyeq B$ means that inf $A \leqslant$ inf B? ($\leqslant$ is the usual order relation in ω, and inf is taken with respect to this relation.)

(Solution is given on p. 36.)

SAQ 22

Let $(X, \leqslant)$ be a partially ordered set. Show that

(i)   each subset of X is partially ordered;

(ii)   for any collection $\mathcal{C}$ of subsets of X

$$\bigcup \mathcal{C} \text{ and } \bigcap \mathcal{C} \text{ are partially ordered,}$$

where the partial order is the restriction of the original partial order $\leqslant$ to the relevant set.

(Solution is given on p. 36.)

SAQ 23

Let $\mathcal{C}$ be a non-empty collection of subsets of a partially ordered set X, each of which is *totally* ordered with respect to the restriction of the order relation on X.

(i)   Prove that $\bigcap \mathcal{C}$ is totally ordered;

(ii)   show, by means of a counter-example, that $\bigcup \mathcal{C}$ is not necessarily totally ordered,

in each case with respect to the restriction of the order relation on X to the relevant set.

(Solution is given on p. 36.)

SAQ 24

Let $\mathcal{P}(X)$ be the collection of subsets of a given finite set X, and for $E, F \in \mathcal{P}(X)$ write $E \preccurlyeq F$ if $\#(E) \leqslant \#(F)$.

Is the relation $\preccurlyeq$ a partial order on $\mathcal{P}(X)$?

(Solution is given on p. 36.)

SAQ 25

Prove that a *totally* ordered set with a minimal element has a unique least element.

(Solution is given on p. 37.)

3.5 THE AXIOM OF CHOICE

3.5.0 Introduction

In Section 3.5 we shall consider some rather profound results. These are the Axiom of Choice and its equivalent statements in the form of Zorn's Lemma and the Well Ordering Principle. Although the Axiom of Choice as here stated may seem rather obvious and trivial, it cannot be deduced from the axioms we have considered earlier, and so it must be inserted as an axiom, if use of it is intended. Certainly its consequences are neither trivial nor obvious, as we shall see on reading parts of the three sections of *Halmos: Sections 15 to 17* inclusive.

At this point we should like to emphasize that these sections should be read primarily for interest and pleasure. You will be expected to recognize the statements of the Axiom of Choice, Zorn's Lemma, and the Well Ordering Principle, but you will *not* be expected to be able to reconstruct the proofs of their equivalence.

In the spirit of the preceding paragraph, there will be no Self-assessment Questions in this section; further, the reading passages will simply be selections from Sections 15 to 17 inclusive of *Halmos*. However, we hope that your interest will be aroused and that you will want to look at the omitted passages.

3.5.1 The Axiom of Choice

READ *Halmos: page 59, Section 15, line 1 to page 61, line 6.*

Notes

(i) *Halmos: page 60, line − 14.*
An informal argument, which can easily be formalized, showing that every relation includes a function with the same domain, runs as follows. If the relation R has a domain X and range Y, then R may not be a function because for $x \in X$ there may be more than one y such that xRy. Indeed, for given x, if Y is not finite, the set $R_x = \{y \in Y : (x, y) \in R\}$ may not be finite; also if X is not finite, the set $\{R_x : x \in X\}$ may not be finite: it may be an infinite collection of infinite sets. Nevertheless, by the Axiom of Choice, we can choose one y_0 from each R_x and the resulting set of pairs of the form (x, y_0) is a function with domain X.

(ii) *Halmos: page 60, line − 13.*
Informally, for each C in $\mathcal{C}$, choose $c \in C$. Then the set of c's so chosen is the required set A.

(iii) *Halmos: page 61, line 5.*
In fact, for our present purposes it is unnecessary to take a longer look!

3.5.2   Zorn's Lemma

*READ **Halmos**: page 62, Section 16, line 1 to page 63, line 4.*

3.5.3   Well Ordering

*READ **Halmos**: page 66, Section 17, line 1 to page 67, line −9, and page 68, line −7 to page 69, line 7.*

General Comment

You may have found Sections 3.5.1, 3.5.2 and 3.5.3 difficult on a first reading. Do not be dismayed; these considerations have occupied eminent mathematicians for a long time and you are not expected to assimilate all the concepts involved. It is necessary, however, to have some appreciation of the concepts introduced, in order to understand the care sometimes required as the mathematics we study gets more sophisticated. For example, it is perhaps surprising to see how an assumption as innocuous as that of the Axiom of Choice can lead, via the application of Zorn's Lemma (we actually omitted the proof from the reading selections), to the fact that *every* set can be well ordered! It is perhaps even more surprising that although the Axiom of Choice, Zorn's Lemma and the Well Ordering Theorem seem so different from each other, in fact each implies, and is implied by, each of the others.

This is as far as we intend taking the study of Naive Set Theory in this course. It is perhaps fitting to have ended with a glance at the fascinating topics of Section 3.5, which are clearly a prelude to further interesting studies.

3.6 SUMMARY OF THE TEXT

This text has been mainly concerned with the axiomatic construction of the natural numbers, and the recursive definition of arithmetic operations upon them. Since we have based everything so far on the idea of a set, of necessity we defined natural numbers as sets. Having defined the set ω of natural numbers as the smallest successor set, the Peano Axioms follow as results.

The natural numbers can lead to the various number systems with which you are acquainted: the integers, rationals and real numbers. The construction of these other systems would eventually follow the lines of discussion in *Unit M100 34, Number Systems*, although we do not attempt it here.

We also discussed order in the natural numbers and the use of the natural numbers in counting the elements of finite sets, distinguishing, in passing, between finite and infinite sets. We then discussed the concept of order in a more general context, and concluded by surveying briefly the Axiom of Choice and related principles. These latter topics were only touched upon here, and may be encountered subsequently.

With this unit we conclude our study of Naive Set Theory. By now you will have gathered that the adjective "Naive" is very relative; in one sense at least, the subsequent units of this course may be very much more naive. That is, we hope that, in the first three units, we have indicated sufficient of the formal development to convince you that our subsequent units could be put on an axiomatic set-theoretic foundation, if we so desired. In actual practice, we shall not so desire, but shall revert to the more "normal" style of mathematics in which the set theory foundations, though comfortingly present, are nonetheless sufficiently unobtrusive to prove no hindrance.

In order to see the first three units of this course in their proper perspective, we suggest that you turn once more to the preface of **Halmos**, where the reader is exhorted to "read it, absorb it, and forget it".

Postscript

> Art thou poor, yet hast thou golden slumbers?
> >   Oh sweet content!
> Art thou rich, yet is thy mind perplexed?
> >   Oh, punishment!
> Dost thou laugh to see how fools are vexed
> To add to golden numbers, golden numbers?
> O, sweet content, O, sweet, O, sweet content!
> >   Work apace, apace, apace, apace;
> >   Honest labour bears a lovely face;
> Then hey nonny, nonny; hey nonny, nonny.

Thomas Dekker (1570?–1641?)
Patient Grissil, Act I

3.7 SOLUTIONS TO SELF-ASSESSMENT QUESTIONS

Solution to SAQ 1

The Axiom of Specification tells us that, in order to specify a set by a condition such as $a \neq b$, we must have a set to which to apply the condition (**Halmos**: *page 6*). For example, given that A is a set, then

$$B = \{\{a, b\} \in A : a \neq b\},$$

would define a set B. However, we have no *guarantee* from the Axiom of Specification or anything else that the expression

$$C = \{\{a, b\} : a \neq b\}$$

defines a set. Indeed, if we were to allow C as a set, we would certainly arrive at a contradiction. If C is a set, then so is $\bigcup C$. But clearly everything is a member of $\bigcup C$, and we have shown that no such set is possible.

Solution to SAQ 2

Recalling the definition of set complementation (**Halmos**: *page 17, line 3*), we have

$$3 - 2 = \{a \in 3 : a \notin 2\},$$

where

$$3 = \{0, 1, 2\}, \quad 2 = \{0, 1\} \quad \text{and} \quad 1 = \{0\}.$$

Clearly,

$$3 - 2 = \{2\} \neq \{0\}.$$

Solution to SAQ 3

(i) $A \cup B$ is a successor set, since

    (a) $A \cup B$ is a set;
    (b) $0 \in A \cup B$, because $0 \in A$;
    (c)   if $x \in A \cup B$, then $x \in A$ or $x \in B$.
        In either case, $x^{+} \in A \cup B$, since both A and B are successor sets.

(ii) $A - B$ is *not* a successor set, since $0 \notin A - B$.

(iii)   We are required to show that $\mathcal{P}(\omega)$ is *not* a successor set.

Since $\mathcal{P}(\omega)$ is a set and since $0 \in \mathcal{P}(\omega)$, because $\varnothing$ is an element of the power set of every set, it is clear that we must find a set $x \in \mathcal{P}(\omega)$ such that $x^{+} \notin \mathcal{P}(\omega)$.

How would it be possible to find a set $x^{+} = y$ such that $y \notin \mathcal{P}(\omega)$? Well, by the definition of $\mathcal{P}(\omega)$,

    if $y \in \mathcal{P}(\omega)$, then $\alpha \in y \Rightarrow \alpha \in \omega$.

Therefore if we can find a

    $y = x^{+}$ such that $\alpha \in y$ and $\alpha \notin \omega$,

then $y \notin \mathcal{P}(\omega)$, and the result will be proved.

PROOF
Since $1 \in \omega$, $\{1\} \subset \omega$, and so

$$\{1\} \in \mathcal{P}(\omega).$$

Let us choose $\{1\}$ as our set x.

Then

$$y = x^{+} = x \cup \{x\} = \{1\} \cup \{\{1\}\}.$$

Now take as our α of the previous remarks the set $\{1\}$; then

$$\{1\} \in y.$$

But cleariy

$$\{1\} \notin \omega,$$

because *every* element n of ω, with the exception of $\varnothing$, has the property that $\varnothing \in n$, but $\varnothing \notin \{1\}$, since $1 = \{\varnothing\}$ so that $\{1\} = \{\{\varnothing\}\}$.

Therefore

$$y \notin \mathcal{P}(\omega);$$

and so we have found $x \in \mathcal{P}(\omega)$ such that $x^+ \notin \mathcal{P}(\omega)$ and therefore $\mathcal{P}(\omega)$ is *not* a successor set.

Solution to SAQ 4

A	$f(u(n))$	B	$\omega \times X$	C	$(n^+, f(x))$
D	$\omega \times X$	E	$(n^+, f(x))$	F	u
G	$(n^+, f(x))$	H	y	I	u
J	n^+	K	(n, x)	L	$(n^+, f(x))$
M	$f(x)$	N	$(m^+, f(t))$	O	n^+
P	u	Q	x	R	S

Solution to SAQ 5

By a result of **Halmos**: *page 47, line 19*, for all $n \in \omega$, $n \notin n$. Now $n^+ = n \cup \{n\}$, so that $n \in n^+$.

Therefore $n \neq n^+$.

Solution to SAQ 6

Consider the set

$$S = \{n \in \omega : n = m^+ \text{ for some } m \in \omega, \text{ or } n = 0\}.$$

(i) $0 \in S$.

(ii) Suppose that $n \in S$ and $n \neq 0$; this implies that $\exists m \in \omega$ such that $n = m^+$.

Then $n^+ = (m^+)^+$, where $m^+ \in \omega$ because $m \in \omega$, so $n^+ \in S$.

(i) and (ii) show that $S = \omega$.

This result shows that, if $n \neq 0$, then $n = m^+$ for some $m \in \omega$.

Solution to SAQ 7

Let E be a subset of $n = \{0, 1, 2, \ldots, (n-1)\}$. If $0 \in E$, then 0 is the element k in question, for $0 \in m$ whenever m is an element of E distinct from 0. If $0 \notin E$, then if $1 \in E$, 1 is the relevant element. This argument may be continued until the (finite) set E is exhausted, and eventually we must find the required element k.

This involves a special case of the Well Ordering Principle for the natural numbers, discussed and proved in Section 3.3.2, Note (vii). The above, deliberately naive, proof is perfectly acceptable for finite sets, but is clearly inapplicable in the general case: when the set E is not finite, the process *cannot* be continued until E is exhausted. This is the reason that the proof in the general case is so much more complicated.

Solution to SAQ 8

(i) $u(n) = a$ for all $n \in \omega$.

(ii) For even n, $u(n) = (a_1, a_2)$.
       For odd n, $u(n) = (a_2, a_1)$.

Solutions to SAQs 9, 10 and 11

m	n	$s_m(n)$	$p_m(n)$	$e_m(n)$
0	0	0	0	1
0	1	1	0	0
1	0	1	0	1
1	1	2	1	1
1	2	3	2	1
2	1	3	2	2
2	2	4	4	4

Examples:

$$s_1(1) = s_1(0^+) = (s_1(0))^+ = 1^+ = 2$$
$$\begin{aligned} p_1(2) = p_1(1^+) &= p_1(1) + 1 \\ &= p_1(0^+) + 1 \\ &= (p_1(0) + 1) + 1 \\ &= 1 + 1 = 2 \end{aligned}$$
$$e_2(1) = e_2(0^+) = e_2(0) \times 2 = 1 \times 2 = 2.$$

Since

$$e_1(2) = 1 \neq e_2(1),$$

exponentiation is not commutative.

To show the non-associativity of exponentiation, writing for simplicity

$$e_m(n) = m \circ n,$$

we must show that there are values k, m and n such that

$$(k \circ m) \circ n \neq k \circ (m \circ n).$$

Choose

$$k = 2, m = 1, n = 0;$$

then

$$(2 \circ 1) \circ 0 = 2 \circ 0 = 1$$

but

$$2 \circ (1 \circ 0) = 2 \circ 1 = 2;$$

therefore exponentiation is not associative.

For sums and products, commutativity and associativity may be verified from the table. For example, $s_1(2) = s_2(1) = 3$, and $p_1(2) = p_2(1) = 2$.

Solution to SAQs 12 and 13

We have the following dictionary:

Recursion Theorem	Sum Function	*Product Function	†Power Function
$a \in X$	$m \in \omega$	$0 \in \omega$	$1 \in \omega$
$f: X \longrightarrow X$	$+: \omega \longrightarrow \omega$	$s_m: \omega \longrightarrow \omega$	$p_m: \omega \longrightarrow \omega$
$u: \omega \longrightarrow X$	$s_m: \omega \longrightarrow \omega$	$p_m: \omega \longrightarrow \omega$	$e_m: \omega \longrightarrow \omega$
$u(0) = a$	$s_m(0) = m$	$p_m(0) = 0$	$e_m(0) = 1$
$u(n^+) = f(u(n))$	$s_m(n^+) = (s_m(n))^+$	$p_m(n^+) = m + p_m(n)$	$e_m(n^+) = m \times e_m(n)$
		$(= s_m(p_m(n)))$	$(= p_m(e_m(n)))$

* We have modified the defining relation of **Halmos**: *page 50, line* -3 to $p_m(n^+) = m + p_m(n)$ in order to agree with the Recursion Theorem.

† We have similarly modified **Halmos**: *page 51, line 7* to $e_m(n^+) = m \times e_m(n)$ to agree with the Recursion Theorem.

Solution to SAQ 14

We take

$$f : n \longmapsto e_m(n) = m^n \qquad (n \in \omega).$$

Writing u_m for u, we have

$$u_m(0) = 1,$$

and

$$u_m(n^+) = f(u_m(n)) = m^{u_m(n)}.$$

So

$$u_m(1) = m^1 = m, \; u_m(2) = m^m,$$

$$u_m(3) = m^{(m^m)}, \; u_m(4) = m^{(m^{(m^m)})}$$

and

$$u_m(n) \text{ is a ``pyramid'' of } n \; m\text{'s,} \; m^{(m^{(m(\cdots}}$$

which we define to be $m * n$.

We see that, for example,

$$2 * 3 = 16 \text{ and } 3 * 2 = 27$$

so $*$ is not commutative. Further, $*$ is not associative: for instance

$$(2 * 1) * 2 = 2 * 2 = 4$$

and

$$2 * (1 * 2) = 2 * 1 = 2.$$

Solution to SAQ 15

First, by the Well Ordering Principle for the natural numbers, $\min E$ and $\min F$ exist. By definition,

$$\forall x \in F, \min F \leqslant x, \text{ and } \min F \in F.$$

Now

$$\forall y \in E, \text{ since } y \in E \Rightarrow y \in F,$$

$$\min F \leqslant y.$$

In particular, for $y = \min E$,

$$\min F \leqslant \min E.$$

Solution to SAQ 16

We verify diagrammatically that the relation of equivalence between subsets of a set X is an equivalence relation in $\mathcal{P}(X)$.

Let $E, F, G \in \mathcal{P}(X)$;

(i) $\sim$ is reflexive:

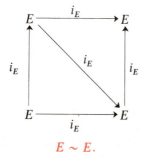

$$E \sim E.$$

The roles of f and f^{-1} of Section 3.3.3 are played by i_E, the identity map in E.

(ii) $\sim$ is symmetric:

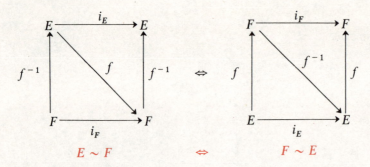

$$E \sim F \qquad \Leftrightarrow \qquad F \sim E$$

(iii) $\sim$ is transitive:

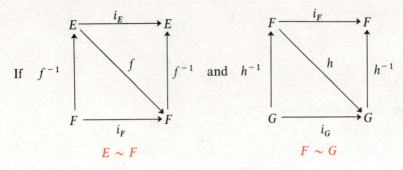

$$E \sim F \qquad\qquad F \sim G$$

then, by combining the above diagrams, we obtain

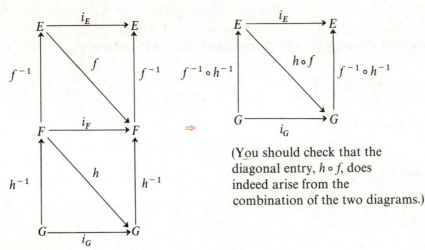

(You should check that the diagonal entry, $h \circ f$, does indeed arise from the combination of the two diagrams.)

We see that

$$E \sim G,$$

using the one-to-one and onto functions

$$h \circ f : E \longrightarrow G,$$
$$f^{-1} \circ h^{-1} : G \longrightarrow E.$$

The above results show that the relation $\sim$ is an equivalence relation in $\mathcal{P}(X)$. (A non-diagrammatic solution is also acceptable.)

Solution to SAQ 17

Every subset of a finite set is finite (**Halmos**: *page 53, line 15*). Since $(E - F) \subset E$, and E is finite, $E - F$ is finite. Further, $E \cup F = (E - F) \cup F$, where $(E - F) \cap F = \varnothing$, and so $E \cup F$ is the union of two disjoint, finite sets. Therefore by **Halmos**: *page 53, line* -10,

$$\#(E \cup F) = \#((E - F) \cup F) = \#(E - F) + \#(F).$$

Solution to SAQ 18

Let $F \sim n$. That is, there exists a one-to-one, onto function $f:F \longrightarrow n$.
Then $f|E$ is a one-to-one function from E onto a subset S of n. As each subset of a natural number is equivalent to a smaller natural number, (***Halmos***: *page 52, line 18*), S is equivalent to some m, where $m \leqslant n$. Since $E \sim S$ and $S \sim m$, $E \sim m$ (transitivity of $\sim$). Therefore

$$\#(E) = m \leqslant n = \#(F).$$

Solution to SAQ 19

(i)   Yes.

(ii)

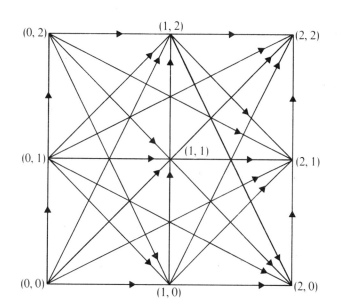

(iii) Yes, $(0, 0)$.

(iv) Yes, $(2, 2)$.

(v)   Yes.

(vi) No.

(vii) The least element is $(0, 0)$: there is no greatest element.

Solution to SAQ 20

We show that $(\mathcal{C}, \subset)$ is a partially ordered set, where $\mathcal{C}$ is the collection of all non-empty subsets of a given non-empty set X.

$\subset$ is an order relation on $\mathcal{C}$, because:

(i) $\subset$ is reflexive on $\mathcal{C}$; $A \subset A$, for every $A \in \mathcal{C}$;

(ii) $\subset$ is anti-symmetric on $\mathcal{C}$; $A \subset B$ and $B \subset A \Rightarrow A = B$, for $A, B \in \mathcal{C}$;

(iii) $\subset$ is transitive on $\mathcal{C}$; for if $A \subset B$ and $B \subset C$, then $A \subset C$, for $A, B, C \in \mathcal{C}$.

(i), (ii), (iii) imply that $\subset$ is a (partial) order relation on $\mathcal{C}$.

Each singleton is a minimal element of $\mathcal{C}$, for if $x \in X$, $\{x\}$ is such a singleton, and

$$A \in \mathcal{C} \text{ and } A \subset \{x\} \Rightarrow A = \{x\}, \text{ since } \varnothing \notin \mathcal{C}.$$

Therefore there is no element of $\mathcal{C}$ *strictly* smaller than $\{x\}$.

If a least element exists, it is clearly a singleton. Suppose that X contains at least two distinct elements, x and y. Then $\{x\}$ is not a least element of $\mathcal{C}$, since it is *not* true that $\{x\} \subset \{y\}$. Similarly, $\{y\}$ is not a least element of $\mathcal{C}$. Therefore no singleton is a least element, and so there is no least element (unless X is a singleton).

Solution to SAQ 21

No. The relation $\preccurlyeq$ is not an order relation because it is not anti-symmetric. For example, the sets

$$A = \{2, 3, 7\} \text{ and } B = \{2, 4, 6\}$$

are such that

$$\inf A = \inf B = 2,$$

so that

$$A \preccurlyeq B \text{ and } B \preccurlyeq A,$$

but

$$A \neq B.$$

(However, $\preccurlyeq$ is reflexive and transitive.)

Solution to SAQ 22

$(X, \leqslant)$ is a partially ordered set.

(i)   Let $E \subset X$ be a subset. Clearly, when restricted to the elements of E, $\leqslant$ retains its properties of being reflexive, anti-symmetric and transitive, and so its restriction defines a partial order on E.

(ii) Since $\bigcup \mathcal{C}$ and $\bigcap \mathcal{C}$ are subsets of X, the remark of (i) above is applicable, and they are partially ordered with respect to the relevant restrictions.

Solution to SAQ 23

(i)   By SAQ 22(ii), $\bigcap \mathcal{C}$ is partially ordered. Let $x, y \in \bigcap \mathcal{C}$, then both x and y belong to each set of the collection $\mathcal{C}$, and if $x \leqslant y$ in one set then $x \leqslant y$ in each set or if $y \leqslant x$ in one set then $y \leqslant x$ in each set, as each set is totally ordered, and therefore $\bigcap \mathcal{C}$ is totally ordered.

(ii) If X is not totally ordered, there exist elements $x, y \in X$ such that neither $x \leqslant y$ nor $y \leqslant x$ holds. Then each set $\{x\}$ and $\{y\}$ is trivially totally ordered, but if these sets belong to $\mathcal{C}$, then $\bigcup \mathcal{C}$ is not totally ordered.

Solution to SAQ 24

The relation $\leqslant$ is *not* a partial order, for if

$$E \preccurlyeq F \text{ and } F \preccurlyeq E$$

we may deduce that

$$\#(E) \leqslant \#(F) \quad \text{and} \quad \#(F) \leqslant \#(E),$$

whence

$$\#(E) = \#(F),$$

so that

$$E \sim F,$$

but not necessarily that $E = F$. (For example, consider $E = \{1, 2, 3\}$ and $F = \{1, 2, 4\}$.) However, transitivity and reflexivity hold.

Solution to SAQ 25

Let X be a totally ordered set with a minimal element, so that the set

$$S = \{a \in X : a \text{ is a minimal element of } X\}$$

is non-empty. Let $a \in S$, $b \in S$. Since X is totally ordered, either $a \leqslant b$ or $b \leqslant a$. Since a is minimal,

$$b \leqslant a \Rightarrow a = b;$$

since b is minimal,

$$a \leqslant b \Rightarrow a = b.$$

Hence $a = b$. Thus S is a singleton $\{a\}$.

For any $x \in X$, either $a \leqslant x$ or $x \leqslant a$; if $x \leqslant a$ then $x = a$ by the minimal property of a, so, for all $x \in X$, $a \leqslant x$. Thus a is the (unique) least element.

Notes

38

Notes

TOPICS IN PURE MATHEMATICS

1	S	Set Axioms
2	S	Set Constructions
3	S	Sets and Numbers
4	A	Group Axioms
5	A	Group Morphisms
6		NO TEXT
7	T	Metric Space Axioms
8	T	Continuity and Equivalence
9	C	Finite State Machines
10		NO TEXT
11	A	Automorphism Groups
12	A	Group Structure
13	T	Topology Axioms
14	T	Topological Closure
15	T	Induced Topologies
16	C	Turing Machines
17	A	Rings and Ideals
18		NO TEXT
19	A	Special Rings
20	N	Categories
21	C	Recursive Functions
22	N	Universal Mappings
23	A	Euclidean Rings
24	A	Polynomials
25	C	Theory of Proofs
26		NO TEXT
27	T	Connectedness
28	T	Compactness
29	T	Fundamental Groups
30	T	Fixed Point Theorems
31	A	Field Extensions
32	A	Splitting Fields
33	A	Galois Theory
34	A	The Galois Correspondence

The letter after each unit number indicates the textbook required for that unit.

Key	S	Halmos	A	Herstein	N	not based on one of the texts.
	T	Mendelson	C	Minsky		